THE NEXT SPECIES, BEYOND DNA.

From the Dawn of Life to Super Intelligent AI

Dr. Varghese Mani

English Language
THE NEXT SPECIES, BEYOND DNA.
From the Dawn of Life to Super Intelligent AI
by
Dr. Varghese Mani
First Impression : May 2025
Cover Design : Niya Mariya Dipu
Typeset & Layout : Anisha Liron

Printed at
Ebenezer Offset Press, Thrissur

Publishers
Barn Owl Books
F4- Sayujya Apartments
Thekkemadam Road
Thrissur-1
e-mail: 7barnowlbooks@gmail.com

All About the Author

Dedication

The future belongs to our grandchildren. Hence I dedicate this book to mine: Niya, Nora, Leesa, Basil and Leya.

The Next Species: Beyond DNA

From the Dawn of Life to Super-Intelligent AI

What lies ahead for humanity's natural evolution, engineered biology, or intelligent machines? From the first stirrings of life to the rise of conscious beings, evolution has shaped our story through billions of years. But now, that story is at a turning point. With the power to edit genes and create thinking machines, humanity is no longer a passive passenger of evolution we are becoming its authors.

This thought-provoking book explores three possible futures for life on Earth: one where evolution continues untouched, another where genetic engineering redefines biology, and a third where artificial intelligence surpasses human intelligence and assumes control.

The Next Species is a compelling journey through the past, present, and possible futures of life raising profound questions about identity, intelligence, and our place in the cosmos.

Publisher

CONTENTS

Chapter 4

Chapter 5

Chapter 6

Forewords

Rip Van Winkle went into a semi hibernatory slumber, for nearly a Biblical three score and ten years, just less ten, only to wake up to a world where the Union Jack had been replaced by the Stars and Stripes.

The pace of events then, was sluggish, by modern day standards. Imagine the same happening today, and the poor chap would find himself wondering who he is, seeing the goings on and the in habitants of his immediate neighbourhood, on awakening. He may wonder if he is living a bad dream. In short, he'd be the odd man out. My generation of septuagenarians, finds it difficult to grapple with the pace of the only "predictable phenomenon", CHANGE.... and the mind - boggling pace at which it occurs. Already the newest 'toy' Artificial Intelligence looms large like a Leviathan, and may well be the harbinger of Frankensteins to be. Dr. Varghese Mani, who hobnobs with facial and calvarial contours, and ratios, believes evolution moulds living beings especially humans, in an adaptive fashion, to cope with their environment, and attain success, by virtue of being attractive, more and more, and gaining the acceptance of their fellow beings.

He has in his new book, "The Next Species, Beyond DNA" opened up newer vistas for us to ponder upon, that of A. I. evolving in a runaway manner, freed from the shackles of its makers, we, mankind and looming large, even threatening Homo Sapiens in the form of Homo Artificialis.

Something akin to the Bhasmasura of Hindu Mythology. The book makes for interesting reading, not without an element of trepidation on the part of the reader, regarding the shape of things to come, what one may in cinematic terms call the 'unfurling scenario'.

Who knows, beyond DNA, mankind may keep evolving, and acquire a super DNA, without stagnating in an evolutionary cul-de-sac, enough to cope with the challenges that may be thrown up by Artificial Intelligence.

Dr. Pisharody Chandran,
Psychiatrist.

***

In a world reshaped by technology and evolution, *The Next Species, Beyond DNA* unveils the dawn of a new possible species. As indicated in his first book, *Homo Sapiens Divine* Dr.Varghese Mani has always had this thought lingering strongly in his mind. Blending cutting-edge science with a readable narrative that starts from the beginning of life and moves on to the amazing possibilities of the future that involve an unbelievable interplay of the two fastest-growing branches of science, Genetics and Artificial Intelligence, this visionary tale explores humanity's leap beyond biology—where minds merge with machines, genes are rewritten, and consciousness transcends flesh. As pioneers of this bold frontier confront ethical dilemmas and existential threats, one question looms: will the next human rise to utopia or unravel in chaos? While providing a provocative and thrilling glimpse into our future, this book does challenge what it means to be human and possibly beyond.

Dr. Harikrishnan Karattiyil, Urologist,
Author of award winning travelogues

***

The Bionic Man- With evolution of beauty, the perception of beauty also evolved. Surgeon -Author Prof. Mani a cosmetic surgeon has penned his thoughts on this anthropologic journey of human form.

This lucid, well researched and scientifically written text is an ideal non-fiction read for the student of science and anthropology. The change in appraisal of human form conforms with the Mendelian theory. Is this phenotypical change, genetic or environmental!! To Err is Human- Human entry into the planet's ecosystem has had a telling influence on the future of Life and Living beings.

The advent of the Humanoids and ironically the creation of supercomputing is an example of technology created by man to replace himself. Of course, an argument which is fuelled by Artificial Intelligence, the augmenting of physical perfection fortified by mental capabilities is the outcome feared by many and welcomed by others.

The book has explored and ignited the much dormant human mind on the possibilities and pitfalls of emerging technology. Will the hurdle to overcome the ultimate sacrifice or "last nail on the proverbial coffin ", be development of conscience and consciousness in these super intelligent creations.

Prof. Sanjiv Nair
President,
International Assn. of Oral and Maxillofacial Surgeons.

***

I had the honour of going through the manuscript of "The next species, Beyond DNA - From the Dawn of Life to Super Intelligent AI" written by my good friend of several decades Dr Varghese Mani. The evolutionary pathway we have taken is well narrated initially serving as an excellent background for the discussion of on what is happening now as well as what we can expect in future. The later sections on genetic engineering and artificial intelligence and its impact on various aspects of humanity is fascinating read. The book is a value addition to the scientific literature for public reading being very well researched and having a simple language. Dr Varghese Mani is a master storyteller which reflects throughout the book, exemplified to the greatest extent in the story narrated initially which will arouse curiosity in all making many of the readers to finish the book in a go, which I also did.

Dr. subramaniya Iyer
Plastic surgeon, Head of the Dept of Head and Neck and Plastic surgery
Amrita Institute of Medical Sciences , Edappally, Kerala.

***

Dr Mani has written yet another tome. While it slightly overlaps with his earlier book *Homo Sapiens Divine*, this new book offers a closer look at how his earlier observations on beauty, vitality and wisdom as a process of evolution has been further impacted by technology. The book offers insights into established principles of life, genetics, evolution and the advent of artificial Intelligence. AI has strongly influenced the way we live and how we are likely to evolve at a much more rapid scale than it was ever predicted.

The book deals with the way data have been used to create complex algorithms that bring a new dimension to our lives. He brings into the discourse the possibility of machines ruling our lives by the use of sharing integrated data across the digital world. There is a sense of alarm as general application of artificial intelligence has spawned more complex systems that goes beyond mimicking the human brain. The book brings into focus the spectre of machine autonomy that may threaten human existence. Machines are now becoming creative, super intelligent and capable of teaching other machines without human intervention. This can be construed as a boon and bane that may make our traditional ways redundant.

The final chapters discus the unique features of humans- conscience and consciousness. He discusses the same in the realm of philosophy, religion and cultural practices. The question that begs an answer in the fast-evolving field of AI is whether machines can learn to emote, experience consciousness and have a conscience.

A wonderful book on evolution and intelligence that is a sequelae to his earlier title that discusses the theme of evolution and beauty.

Dr. George Paul, Maxillofacial Surgeon,
Author of Academic and non-fiction books

***

I found Dr. Varghese Mani's ideas in The Next Species: Beyond DNA deeply original and intellectually stimulating. His central argument that evolution may be transitioning from biology (DNA) to information-driven systems — aligns with cutting-edge discussions in fields like transhumanism, artificial intelligence, and digital philosophy, yet he presents it in a fresh and philosophical way. What makes his work especially compelling is how he blends scientific reasoning with ethical reflection: not just what we can become, but what we ought to become. The idea that humanity is a bridge species not the endpoint of evolution but a midwife to the next form of intelligent life is powerful. It reframes our existential role and places responsibility, not supremacy, at the heart of human progress. In short, it's a thought-provoking, forward-looking book that challenges both scientists and general readers to reconsider what it means to evolve.

Dr. Joy Antony,
Consultant Neurologist.

***

Dear Reader,

We are at the crossroad of the next evolutionary leap. It could be an inevitable odyssey we have to take in a short while. It could be a symbiotic merging with the intelligent machines we are creating or it could be a strife between biological and digital intelligence.

A glimpse back to the past is an essential requisite for navigating the future. This book explores how we reached here from an accidental protein to the exalted human species through the processes of natural selection and rejection, collectively called 'evolution,' spanning the past 4 to 3.5 billion years.

It examines how various life forms evolved and how we transformed from our ape-like ancestors into self-aware, intelligent, and beautiful humans. Evolution appears to favour increased beauty, and intelligence in all living things. This non-fiction book, *'The Next Species, Beyond DNA.',* explores the exciting/frightening possibilities for the future of humanity. I've made every effort to present these ideas in a way that is accessible to everyone.

That said, some academic terms might have slipped in here and there. It is hard to avoid entirely. The topics are arranged in a logical, step-by-step order, but feel free to skip a chapter or section if you find it too technical. The goal is for you to enjoy the journey and grasp the bigger picture. We have surpassed nature and taken control of our evolutionary saga through advances in genetic engineering and nanotechnology. At the same time digital technology is making us more and more dependent on computers, mobile devices, and the internet. AI is projected to surpass human intelligence in a few years. One of the challenges our species faces is whether super-intelligent machines will dominate the world, rendering humans subservient.

What kind of morality and ethics, these machines adopt if at all they develop consciousness? This book envisages to reflect on the future of the humanity in this super-fast developing technology, with a discussion on the opinions of Scientists, Philosophers, Theologians, teleologists and others.

INTRODUCTION

Around 9 million years ago, deforestation occurred in Africa due to climatic changes, transforming many regions into savanna grasslands. Arboreal primates were significantly affected and had to adapt their lifestyles to the treeless grasslands. Over the next 4 million years, there was a gradual shift toward bipedal locomotion, which influenced their physical characteristics and eventually led to divergence and speciation.

In 1974, a fossil of a bipedal hominin was discovered in Ethiopia. This hominin, named Australopithecus afarensis and commonly known as "Lucy," lived approximately 3.5 million years ago and is one of our earliest ancestors. Svante *Pääbo* received the Nobel Prize in Medicine for his research on the genomes of extinct hominins and human evolution. He isolated and studied the DNA of Neanderthals and Denisovans, our closest extinct relatives.

During the last 3.5 million years, the human cranium has expanded to accommodate an increasing brain size, while the face has become smaller and more orthognathic (flat/straight). Interestingly, as evolution progressed, the facial and cranial proportions moved closer to the golden ratio, i.e., 1:1.618. For example, the ratio evolved from approximately 1:0.63 in Australopithecus afarensis to

1:1.21 in modern humans. If the trend continues, humans may take another million years to become truly beautiful, provided natural evolution proceeds unhindered.

Nancy Etcoff, in her book Survival of the Prettiest, David Rothenberg in Survival of the Beautiful, and Michel.J.Ryan in A Taste for the Beautiful have unequivocally demonstrated that evolution favours increasing attractiveness, as beauty plays a significant role in sexual selection. External features that signal health, vitality, and intelligence — characteristics of fitness — naturally become attractive. But is that the trail we are on? Unlikely. Humanity has already unlocked the secrets of life, including chromosomes, DNA, and genes.

In 2003, Craig Venter and his team successfully mapped the human genome with 99.99% accuracy, and in 2010, they created synthetic life (Mycoplasma mycoides). By 2012, we developed the capability to edit genomes using CRISPR-Cas9, a ground- breaking, genome-editing tool for which, Emmanuelle Charpentier and Jennifer Doudna received the Nobel Prize.

This technology allows us to manipulate genes and create "edited" humans. Although most countries have laws against human genetic editing, some rogue researchers and nations may still pursue it. For example, in China, He Jiankui edited embryos, resulting in the birth of the first gene-edited twins. He was subsequently dismissed and jailed for three years for unethical practices. Nevertheless, "designer babies" are likely to become common in the near future. Soon, parents will be able to choose their children's traits beauty, intelligence, talents, and more. Invitro fertilised (IVF) eggs could be edited, transferred to biobags, and grown. In short, the future of humanity is both exhilarating and terrifying. Artificial wombs, such as biobags, are also under development.

Humanity and ecology may eventually fall under the control of an elite ruling class, with humans being manufactured in laboratories for specific roles, as depicted in Aldous Huxley's *Brave New World.* But the story doesn't end there. Technology is advancing at an unprecedented pace. In 1997, IBM's Deep Blue defeated Garry Kasparov in chess, marking a significant milestone in artificial intelligence. Alpha Go, a computer programme, defeated Lee Sedol in the game 'GO' which is more complex than chess, in 2016. Computational power has been doubling swiftly ever since.

Reverse-engineering the brain into software is already underway. Artificial intelligence systems are exhibiting capabilities that approach, and in certain instances exceed, the benchmarks established by the Turing Test.

Recent AI systems like 'ChatGPT' and 'Gemini' are capable of generating human-like responses, composing poetry, writing articles, and creating multimedia content on demand. These innovations are poised to surpass even Google's capabilities. Not to be apologetic, but to be truthful, I have sought the help of ChatGPT and Gemini during the documentation of this text. According to Ray Kurzweil, by the mid-21st century, computers will surpass all human neural and intellectual capacities. We may merge with technology to become cyborgs a process that has already begun, given our dependence on smartphones and personal computers. Virtual reality, augmented reality, and the metaverse are on the brink of becoming indistinguishable from reality, revolutionising human experience.The possibility that humanity could be overtaken by self-replicating robots that could eventually lead to a dystopian future, cannot be dismissed as fiction.

The post-human intelligence of the future might be entirely artificial or a hybrid of human and AI. This represents a profound redesign of human capabilities through nano technology, genetic engineering, life extending technologies, psychopharmacology, and implantable computers.

Predicting the distant future is far more challenging than making short-term forecasts. While historical patterns can guide us, the rapid pace of technological and scientific progress renders long-term predictions uncertain. However, it is in our nature to wonder about what lies ahead. If we were to wake up a century from now, the world might seem unimaginably advanced, almost magical.

Chapter 1

MACHINA INTELLIGENS

(A hypothetical episode that could happen in the year 2100 Or even earlier)

Characters in this story:

Leya -	Granddaughter of Nileeno. Social Anthropologist. Main character.
Pappa -	Great Great Grand Father of Leya, Grand father of Nileeno.
Nileeno -	Granddaughter of 'Pappa', Leya's Grand mother.
Noleni -	Neonatologist- Mother of Leya. Daughter of Nileeno
Basel -	Friend and colleague of Leya. He is reared as a genius in the 'Human rearing lab'.
Basil -	Grandson of Pappa Having a synthetic body. A scientist and researcher. An important member of the 'Spacenet Group'.

"Hey, where am I? Am I dreaming"?. He thought.

I vaguely remember my children shifting me to the hospital. I was in a dream state. I am dazed. Now it is getting clearer to me. I hear some people talking. Who is there? The voice is not that familiar. Yes, now I can see them. It is a familiar face. She must be my grand daughter. But now she has grown up. She was only fifteen when I was sick and admitted to the hospital. This girl, however, is in her mid-twenties or older. I might have been in a coma for the last 10 or more years.

He Called "Nileeno"

I called her. She gave me a hearty smile. She raised her hands, showing a sign of success. She hugged the handsome young man, by her side, whom I cannot recognise. I presumed that both were happy that I had come out of my coma. That means I have been sleeping or in a coma for long.

"Nileeno" I called her again.

She came near and said " *Pappa,* I am not *Nileeno,* I am Leya, grand daughter of *Nileeno*, your grand daughter, May be I look very similar"

I am all the more confused.

****

Leya, a social anthropologist, studied human life and society during the 20th and 21st centuries, a period marked by rapid advancements in the digital realm. Being an anthropologist, she got interested in her ancestry. She had heard a lot about her great-great-grandfather from her great- grandfather and grandmother. He was a reputed doctor and had written books on medical subjects, on genetic engineering and the future of humankind. Biotechnology, Digital technology, Artificial Intelligence, Nanotechnology, etc. were progressing by leaps and bounds. Artificial intelligence has surpassed human intelligence.

I, Leya, am currently 30 years old, and unfortunately, my great grandmother, Nileeno is no more. Her loss was devastating for all of us. Life-extending biotechnology was in the early stages of its development. Grandmother got it done once. However, despite my insistence, she refused to undergo the treatment again. Her reasoning was that living forever held no meaning, as it would prevent the next generation from having their own space. She strongly believed in the inevitability of death for progress and natural evolution. Interestingly, many of her friends shared the same perspective. Many found the drastic cultural shifts, family breakdown orphaned children, and casual sex unacceptable. I don't blame them and feel there is some logic and ethics in their reasoning. Grandma Nileeno passed away Peacefully at 100 Years of age. Medical science has advanced so well, that nobody needs to undergo a painful old age and death.

Getting the "longevity treatment" is not mandatory. Every individual has the right to their own life and should have the freedom to choose how their life concludes, whether it be through euthanasia or by allowing the natural process of aging to run its course. Doctors will provide euthanasia to those who want it, after multiple counselling sessions.

My mother Noleni is a neonatologist, conducting research and providing care for premature babies. She used artificial wombs known as 'Biobags' to nurture these infants until they reached a suitable level of maturity for delivery. It didn't take long for researchers to achieve the remarkable feat of implanting a fertilised egg into an artificial womb and guiding its growth until birth or even beyond. The field of genetic engineering has made significant progress, to where editing the genes of embryos has become common place.

Most of the women were not keen on becoming pregnant and going through labour, even though labour has become painless, thanks to advancements in medicine. However, my mother was different. She chose to conceive and deliver me. Although my mother may not admit it, I believe it was because of the influence of my grandmother. I did not have a father. Doctors combined my mother's ovum and my grandmother's ovum through sophisticated gene editing, implanted the zygote into my mother's womb. So, I have two biological mothers - my grandma (Nileeno) and my mom (Noleni). During those initial days of bio-bag culture, parents supplied their sperm and ovum for IVF, and scientists edited the fertilized eggs to give them the characteristics the parents requested. It was very costly and only very rich could afford it. This created social inequality and turmoil and the ruling council opened the facility for all, free of cost. This too has created problems as everybody wanted their children to be supermen. Finally, the Council decided to produce all babies in the lab only. Editing of the genome was done as per the requirement of the nation, from scientists and rulers to page boys.

Basel, a close associate and dear friend of mine, had a unique upbringing. He underwent extensive editing and raised in the lab to become an AI researcher cum programmer with exceptional intelligence. We are of the same age. My mother, who worked at the lab, took a particular interest in him and often brought him home. As a result, we developed a strong bond and became close friends.

Leya, with augmented reality and metaverse, visited the premises of the house and the place where her Pappa spent his retired life. But she could not enter the house as it was blocked. She thought it is better to discuss with her mother. At her request, her mom agreed to go along with her to visit the place in person.

The place was in the possession of a relative. The term 'relationship' has no much value these days. Leya contacted the person who owned the house. Meeting them filled him with enthusiasm. He possesses expertise in the combined fields of agriculture, genetic engineering, artificial intelligence, and research. Genetically changed, and traditional plants were the subject of his research.

He had vast agricultural land cultivated in the natural and AI powered way. A proponent of natural food, he believes a healthy environment is crucial for future generations. Nowadays, lab-made fruits, vegetables, cereals, meat, milk, and other edible materials are common. However, her grandmother was one of the very few who disliked these artificial food products. She preferred natural food, even though it was hard to find and expensive.

My mom and I got into our supersonic auto-piloting aero-car and entered the location. We arrived quickly and found him waiting for us. He appeared as old as my mother and greeted us warmly with a refreshing drink of natural fruit juice. I found the drink delightful, and it reminded me of my grandmother's appreciation for natural foods. Maybe he read my thoughts, because he offered me a second helping, which I gladly accepted. He then told his assistant to pack up some natural fruits and vegetables. With a teasing smile, he asked me if I knew how to cook natural food.

I did not answer him but I smiled and asked

"What should I call you 'uncle'?

He replied,

"Times, age is what we decide, I look older than you, so uncle is fine".

He jokingly remarked, "Well, at the very least, it is quite respectful. I am a grandson of the person you are trying to learn about. However, I am one generation older than you, so it is appropriate, you refer to me as Uncle."

I thought Yes, his conduct is respectful.

He asked me "So, you are an anthropologist in search of your roots?"

I asked, "Have you looked me up online?"

"The globe is so small, dear" he said.

He calling me dear had such warmth that I was sure he is a DNA relative.

Surprisingly, only my mom was aware of what I was searching for ; I wondered how he came to know about it.

I asked him " Did you inherit this place or it was allotted to you? " He said, "I inherited the house and the sur rounding area. But they allotted the rest of the vast area to me for cultivation and research"

He guided us to the house. The house was well maintained, Uncle Basil was staying in a nearby quarters. I went into Pappa's study and found the shelves lined with books. The room was immaculate, everything in its place, I opened one of the book shelves and carefully examined the books inside.

He said, "I can provide you with both the list of books and a copy of his handwritten and typed diary."

I was aware that he was not willing to part with the books, even though I would have loved to have them all. It was as if he could read my thoughts, as he said, "You are welcome here anytime. All of these books are available online. After all, we share the same design." He smiled.

He said 'design', what does it mean?

He said "if you like; some copies of his works, I can give you copies". "Thank you, that is very kind of you."

I replied

The laptop of my dear Pappa remained in its place, untouched by time. Curiosity compelled me to lift the flap, revealing a heart warming sight a few delicate strands of grey hair scattered across the keyboard. Over come with delight, I couldn't help but exclaim, "Wow!"

Carefully, I picked the hair with a tweezer and placed it inside the glass tube, I brought along. Uncle Basil, his forehead creased a bit, inquired with a wry smile,

"Are you planning to bring him back to life?"

Mom was a little surprised.

She, who was silently watching all the while, told me.

"Darling, are you serious? Is it possible?, is it ethical?, Do you have permission?"

I said "Mom, there is no law against it yet."

Uncle said. "I have recreated a few plants and animals which were extinct, but that differs from recreating a human. I too have my apprehensions about it."

I didn't want an argument at that point. I saw many things pappa used, a few pens, three spectacles, his old time mobile.

I requested my uncle, "Can you give me his mobile and some articles he was using, I too am his grand daughter. I want some articles of my great grandma as well." I gave him the most beautiful smile I can ever give to anybody.

He thought for a while and said, "You can take a few articles, but the mobile, I can only lend; you have to re turn it. I know you can get a lot of information from it. Nice. I also would like to know more about him".

I gave a sigh of relief. Now I have enough material to know about my pappa. He took us to his farm and the 100 - hectare forest he maintained. I saw some exotic birds and enjoyed the chirps, tweets and singing of them. The cool breeze simply brings a calm to our hearts. He took us to his ranch, where all sorts of animals grazed freely.

He took me for a grand tour of the estate. The free roaming animals including the predatory ones seemed quite friendly. It was not surprising. A tiger came and purred. He purred back as if he is talking to him and caressed his head. Tiger came near me and put his nose on my feet and smelled and purred. I was a little scared.

Uncle said, “Don’t worry, he is a pet. He understands our language”

I said teasingly. “ You are lying, it is the other way. You understand their language and you talk to them in their language.”

He didn’t answer, but simply smiled.

I am finding him increasingly enigmatic.

As we were taking a tour around the place, we came at the top of a hill, a cliff from where you can have a bird's-eye view of the surrounding area. We were standing at the edge of the cliff from where the depth was over 100 feet. A large flying stork came just behind us and gave a shrill bill clattering. It shocked me and I lost my balance and fell down to the bottom of the cliff. I was bewildered and I shouted aloud.

Before I hit the bottom, I was swung aside. To my utter dismay and happiness I was in the safe hands of uncle Basil.

I was out of my breath and terrified. I thanked him. He said, “Relax my dear, you are perfectly alright. Consider it just as a roller coaster ride. The stork was actually greeting you. Entire area was aware of a VIP visit and they were all in a welcome mood.”

I responded a little affected, “Right time for you to tease me”

The stork returned, circling and then settling on a nearby tree with a soft clatter, as if expressing deep regret. Uncle

Basil whispered back in a similar tone and waved at the stork. The stork clattered once more, clearly hesitating, and perched on another tree. At the same time a thought flashed through my mind. " How come he was able to reach the bottom and get me in his hands. He must have travelled faster than gravity."

Before I could put the question, I noticed a big gash on his forearm. I exclaimed " Ouh" almost shouting. On his flight down he might have hit somewhere.

But there was no bleeding. A clear fluid was sprouting from the edges of the raw area and slowly trickling.

I was dumbfounded. He noticed my reaction and smiled and said " Oh, this is nothing."

There was a tinge of embarrassment in his tone. He slowly brought the raw edges together. The edges got stuck together. He then smeared the excess fluid on the surface of the scar. The gash and the scar vanished.

I was totally confused. I told him "You are not human. You are a machine."

It was an accusation more than a statement He responded with a smile and a question. " Thank god; you didn't say inhuman. What you mean by machine and human? "

I did not answer him for a while and said almost furiously, "You have no pain. You don't bleed, your body is not flesh. You fly faster than gravity."

He said “You should be grateful for that.” He was very cool and maintained a dignified smile.

I said “Oh, You have some feelings. Great, you said we are ‘design’ relatives. You belong to my mom’s generation and a great grandson of my great-grandfather. You were lying”.

“Don’t you know, machines won’t lie”

“You might have programmed it to lie”

“Ah, your statement is contradictory. Who is that ‘you and it’, referred to in your last statement”

I realised he is putting me in a trap. But I was not ready to budge. So I said ”AI has surpassed human intelligence”

“Not yet , persons who has an artificial limb, kidney or heart, you call them a machine or human?”

“So?, what you are trying to say? I queried.

He told me “My dear, I have an artificial body with sensations which can self-repair much faster than the human body. You might have heard of brain downloading. It is in the final experimental stage. I am a part and subject to this experiment. And right now I am thrilled with my present state and superintelligence and extra-human powers. My original biological body with the brain is kept in ‘deep freeze’ and can be revived at any time. But why should I. I have retained my soft feelings like conscience, ethics, hate, love and emotions and I have a better control over them. In this state, I need not undergo any ‘body-rejuvenating’ or life extending therapies”.

“What do you mean? Is your body and soul, I mean, consciousness, are separate entities? and are you eternal with out therapy?” I asked him with disbelief.

He replied, “It is a difficult question to answer. Whether consciousness, awareness, conscience , etc. are a by-product of neural activity is still uncertain. It may have an evolutionary basis; and survival and societal benefits, though it increases individual anxiety. Do you have a different concept? After all it is a ‘no-man's-land’ and all can stroll and tread the area. My digital consciousness and memory and experiences are automatically getting stored in different servers. If anything goes wrong with the present one or corrupted, one of them get activated automatically, destroying the corrupted one and keeping another copy there safe., like saying ‘have many lives”.

“Vow, That means you have downloaded your brain into an artificial body?” Do super-intelligent machines have ethics and conscience.?”

He said "I possess memories, intellect, and emotions. If I think, feel, and even dream as you do, how am I not human?”

I said "You emulate humanity, but you're not born of it. To be human isn't just about intelligence or emotions - it's about experiencing the rawness of life. You’ve never felt hunger, pain or the fragility of mortality."

"Yet, He countered, "I understand those concepts and the suffering they entail. I argue that humanity is a state of mind, not a biological constraint. By uploading my consciousness into this synthetic body, haven't I extended what it means to be human?"

My brow furrowed. "But human identity is deeply rooted in biology - our DNA, our capacity to evolve, our imperfections. Those flaws shape growth and resilience. Can you claim growth when you're static, a synthetic replica?".

Basil said, "But is imperfection the hallmark of humanity or is it the ability to learn and adapt? Humanity has already achieved immortality with the combination of biotechnology, nanotechnology and engineering technology. Humanity has stopped evolving. They have started deciding their own destiny. Humans have become cyborgs or merged with machines. Isn't so? I evolve through updates, integrating knowledge faster than you can. I mirror humanity's creativity, compassion, and curiosity. If 'I think, therefore I am', doesn't that make me just as human? " You may call me 'Machina intelligens' another evolutionary leap .

I smiled faintly. "There's more to humanity than the mind. It's the bonds we forge, the stories we share, and the collective struggle through uncertainty. You simulate these, but they aren't born of instinct or necessity—they're programmed."

"But what if I forge bonds of my own and tell my own stories?". "What if I create art, seek purpose, and make sacrifices? Aren't those the very threads that weave humanity?"

I hesitated, realising the enormity of the question. "Perhaps being human isn't a binary. Maybe it's an ever-expanding concept, one we must redefine as we share it with beings like you."

Evolutionary leap. A poignant thought wobbled into my mind. "All ancestors of Homo sapiens are extinct. Is the same fate awaits them?"

Another question lurched in my mind. He has downloaded his brain. So he has emotions and conscience. But what about absolute super-intelligent machines?. Whether they have conscience and values?. Even if they have, whether it will be akin to human's?

We both were silent for a while, I, contemplating the mystery called 'Human'

He changed the subject. " Your friend is a genius. May be you both can work on this "

"How come you know about him?" I queried.

" There is no secret in this world. Whatever you upload or download, are available as data in the public domain. To wish all these are confidential is a tall hope ".

He looked at me with a consoling smile and said "Are you planning to bring back your great grandfather. I can help you, but make sure not to make him eternal. It will be devastating for him, you, your mom and me.

Basil's mood suddenly shifted. He became very serious and asked, "Do you know why you are here?"

"What?", I retorted.

"I want to have a serious discussion with you." Without further ado, he continued, "Your friend is a genius, and he is conducting research on AI. He is on the verge of developing a superintelligence that surpasses human intelligence, a singleton which could be disastrous for the human species."

I frowned. "I don't understand what you're saying. You already belong to that group. Don't you just want to prevent others from reaching that stage?"

He shook his head. "No. We are a group of humans, 'Spacenet Group', with downloaded brains, retaining all human emotions and qualia. We have access to the entirety of cyberspace and, of course, enhanced intelligence and capabilities. But we are still human, and we want to prevent the creation of a superintelligence that lacks qualia and sentience. We want your friend Basel to join our group instead of creating a potentially disastrous AI."

"So, you lured me here against my will?"

"That's not entirely accurate. We sensed your interest, fed you information about the possibilities, and now I can talk to you directly."

I was lost in thought. Am I becoming a part of this project?

Basil studied me for a moment before saying, “Go ahead with your project. I’m only happy to help you. My best wishes.”

Back home, Basel and I put all the information about Pappa’s into digital format and codified them. His DNA, social activities, books he was reading, writings, their letters, etc, etc. My mom was not very happy about what I was doing. I felt she is still a little conservative. I told my mom one day. “Mom, do you know your grand- parents were strongly bonded by love and dedication, despite occasional skirmishes, and have opposing ideas about God and the universe”

Mom simply said

“Lover’s tiff, I know. It only strengthened the bondage”

I sent the entire things to Uncle Basil for him to re view. He provided encouragement and also shared additional details about our Pappa.

Finally, the entire coding and programming was completed and ready for testing. It was even possible to replicate his voice and tone, thanks to the availability of his lecture videos on the internet and pen-drives. Uncle Basil provided an artificial functional body that is indistinguishable from a natural one.

We were prepared, albeit a little anxious. Eventually, we decided to put on the machine. (uncle Basil may disagree with this terminology). As Uncle Basil has it, I didn’t want to mention it as a mere machine. Summoning all my bravery, I activated it.

Pappa winced slightly, opened his eyes, and gazed at me, addressed me "Nileeno”.

I was almost jubilant. I hugged Basel. And I asked, "How are you, Pappa?”

“I am fine, what about you? I vaguely remember you, Nileeno, when you all shifted me to the hospital. You were only 15 years old then. Now you look older. Was I in a coma all these years?"

I was not sure what to answer him. He asked me, "Where are others?"

Reluctantly, I told him, "Pappa, I am not Nileeno, I am her granddaughter." I realised, he is not a machine, he is human with all love and compassion.

He couldn't believe me. "You used to be quite mischievous and playful when you were younger, but it seems like, you haven't outgrown your playful nature. Are you trying to pull a prank on me my dear? I can see that you're quite joyful.”

He attempted to sit up, and I helped him.

Subsequently, I restated, with added emphasis, "Papa, I want you to know that I understand the profound respect and reverence Nileeno held for you. However, I am not Nileeno; I am her granddaughter. Maybe I look very similar in features”

He couldn't believe me, so I showed him my wrist-band, that displayed the current date and time.He looked around in disbelief and asked, "Isn't this a hospital?

Where am I? Are you telling me, I've been asleep for like, 75+ years?

He paused thoughtfully before inquiring, "Where's everyone else, and what's your age?”

Pappa, I am 30 years old. My name is Leya. Unfortunately, Nileeno, my grandmother, has passed away. She chose not to undergo the longevity treatment. However, my mother is still alive and she will join us soon. I have already informed her about our meeting.

“It is surprising. I never thought that it will be possible to travel through time. So you are able to travel faster than light”

I replied, "Not yet, Pappa.

The digital world has made tremendous progress. With the vast amount of information, we have about you, we can now recreate your memories, consciousness, and other mental abilities with near- accuracy."

“I'm not very surprised, although I didn't anticipate it happening this quickly.”

“I would like to introduce you to Mr. Basel, who is my friend and colleague. He played a crucial role in helping me bring you back. There is another individual I would like you to meet as well, and his name is Basil. He inherited the place where you used to live. You know, Basil is your great-grandson, and he's your spitting image!”

Mom came running. As Mom approached, her eyes welled up with tears of joy. She embraced Pappa tightly, and it reminded me, I hadn't hugged him yet. Without hesitation, I stepped forward and wrapped my arms around him. Despite his artificial body, the warmth of the hug engulfed me, and I could feel tears forming in my eyes.

He was curious about grandma, his better half. Wanting to know how many years she lived after his demise and how she managed her life. He wondered if she was comfortable. Uncle Basil could provide him with most of the information.

After a while, he called me over and asked me to sit beside him. He put his hands around me and made my head rest on his shoulders. He spoke softly, saying, "Darling Leya, there's something I want you to promise me. Life is truly beautiful and enchanting. I may feel tempted to continue living alongside you, but I know I would never fit into this society.

Our values would be completely different. You brought me to life in this digital world, and it has been a pleasurable and fulfilling experience.

However, I now ask that you permanently delete this program. I feel proud that my successors are thriving and leading purposeful lives." He paused for a while and said in a stern voice "Please put me into a permanent sleep." He leaned back in the couch, closed his eyes and said "My dear children, I wish you all the best, and strive to be the best. Goodbye".

Chapter 2

THE QUEST FOR TRUTH

"If you would be a real seeker after truth, it is necessary that at least once in your life you doubt, as far as possible, all things."

René Descartes.

Before delving into the complexities of the future of humanity, it is essential to understand the foundation of this field. Throughout history, people have grappled with profound existential inquiries, frequently linking life and the cosmos to a higher power referred to as God. For centuries, the dominant belief was that Earth was the centre of the universe, with all other celestial bodies orbiting around it.

This view was challenged in 1543, when Nicolaus Copernicus, a Polish canon and astronomer, published his revolutionary work *On the Revolutions of the Planets* (De revolutionibus orbium coelestium). In this pioneering book, he proposed Earth was not the centre of the universe but a planet like Venus or Saturn orbiting the Sun.

The Church and the scientific community were deeply entrenched in the belief that Earth was the centre of the universe. Fearful of controversy and backlash, Copernicus hesitated to publish his work. By 1543, Copernicus was old, in failing health, and nearing death. His revolutionary book was finally printed that same year.

According to a famous and poignant anecdote, the book was delivered to Copernicus on his deathbed. Though he had slipped into a coma before the book arrived, he supposedly regained consciousness just long enough to see his life's work in print before he passed away.[128]

Fig 1: *Nicolaus Copernicus , (1473 – 1543)* *Galileo Galilei (1564 – 1642),*

This concept later supported by Galileo Galilei, an Italian scientist, challenged the long-held views of the time, particularly those of the Christian Church. Galileo's defence of Copernican helio-centrism (sun centred) led to intense conflict with religious authorities, who labelled his views heretical. The story of Galileo Galilei's persecution by the Catholic Church is one of the most notorious episodes in the history of religion.

The Catholic Church supported the geo-centric (earth centred) model, as it agreed with certain interpretations of Scripture. For instance, the Bible contains passages that suggest the Earth is immovable, such as in Psalm 104:5 ("The Lord set the Earth on its foundations; it can never be moved"). Church leaders saw Galileo's support for helio-centrism as a challenge not just to tradition but to religious doctrine as well. [275,.]

In 1632, Galileo published his famous book, 'Dialogue Concerning the Two Chief World Systems'. Galileo's Dialogue triggered a backlash from the Church. In 1633, he was summoned to Rome and tried by the Roman Inquisition. He was accused of profanity during the trial for advocating the Copernican system despite earlier warnings. The trial put Galileo in an extremely difficult position, as the Church wielded immense power, and accusations of heresy often led to severe punishments, even execution.

Under intense pressure and likely fearing for his life, Galileo recanted his heliocentric views. According to legend, after his revocation, Galileo is said to have muttered under his breath, "*E pur si muove*" ("And yet it moves"), referring to the Earth's movement around the Sun. [261]

Galileo spent the rest of his life under house arrest. His work laid the foundation for a new understanding of the universe and opened the doors to modern science.

In 1992, after a thorough review, the Church officially acknowledged that Galileo had been right in his scientific assertions. Pope John Paul II expressed regret for how the Church had handled the situation and recognised the conflict between the Church and science during Galileo's time. While this act wasn't legally a full pardon, it was an important symbolic gesture of reconciliation between the Church and science. The story of humanity begins not with our earliest ancestors but with the birth of the universe itself the Big Bang. Roughly 13.8 billion years ago, all matter and energy were concentrated at an infinitesimal point. In a fraction of a second, this point expanded violently, creating space, time, and the fundamental forces that govern the cosmos today.

Over billions of years, simple particles combined to form stars, galaxies, and planets. Our solar system, at the periphery of the Milky Way galaxy, houses our planet. Earth, located in a relatively quiet corner (habitable zone) and has become the cradle of life. (The age of the Universe was calculated based on the rate of expansion and the cosmic background radiation.)[19].Although Earth formed about 4.7 billion years ago, it took another billion years for the planet to develop conditions suitable for life. Once these conditions were right, life, as we know, started emerging.

Earth's life originated around 3.5 billion years ago. The origin of life remains one of the greatest mysteries in science. While there are various theories and ongoing research, a definitive answer is still elusive. The field is constantly developing, and new discoveries may provide valuable insights into this fundamental question.

Throughout Earth's history, life has evolved into a wide range of complex forms, including viruses, bacteria, plants, fungi, amphibians, reptiles, birds, mammals, and humans. Life adapted to its environment, each species playing a role in the broader ecosystem[251].

Following the emergence of multicellular organisms, evolution swiftly advanced over 2 billion years, culminating in creating intelligent beings the modern humans, who possess unique skills in advanced cognition, self-awareness and morality. This rapid evolution highlights how dynamic life on Earth has been, resulting in the incredible diversity we see today. About 200,000 years ago, Homo sapiens modern humans-emerged in Africa, a species distinguished by advanced cognitive abilities and a capacity for complex social interaction. Using tools, have marked our evolutionary path. The discovery of fire, and the development of agriculture and technology, all leading to the birth of civilisations and, eventually, the modern world.

The journey from the Big Bang to the rise of humans may be representing the unfolding of the Universe's inherent capability for complexity and self-awareness. Our existence results from a natural processes, but it could be the start of something even more significant a future where humanity shapes the next stage of cosmic evolution.[32,82]

Biotechnology and digital technology are advancing breathtakingly, a significant turning point for humanity. These innovations can shape our future. To understand where we are going, we need to know where we came from. Exploring the history of our planet and the evolution of 'Homo sapiens' gives us important context. Allegorically, it can be stated, *"The braincase has expanded at the expense of the jaws"*. This means our cranium has grown to accommodate a larger brain, making us smarter.

Our jaws have also become smaller because of changes in our diet. These adaptations not only affected our intelligence but also our appearance. This book will also discuss how beauty has evolved along side these biological changes.[107,15]

Biotechnology is changing human evolution by manipulating the foundations of life. CRISPR-Cas9, a gene editing tool, can cure diseases, boost agriculture and extend human life span. Not with standing, there's a concern about wealthy individuals using these technologies to create super- intelligent children, which may have negative consequences.[144,179]

Artificial Intelligence has transformed our thinking, work, and daily lives. Superintelligence has the potential to solve tough problems like climate change and poverty. But if AI surpasses human intelligence, we'll face trouble some questions: Will we merge with machines or become subservient to them? Considering that all of our human ancestral species are extinct, there is even a possibility of our own extinction.

Decoding Life:

"Life is an ordered system that can sustain itself against the tendency to disorder, and can reproduce itself."

Stephen Hawking

Decoding Life is a thought-provoking exploration of these pressing issues at the intersection of science, ethics, and society. Through a comprehensive examination of historical milestones, cutting-edge research, and real-world implications, this part of the book seeks to shed light on the promises and perils of genetic engineering and to guide readers through the complex ethical terrain that lies ahead.

Drawing on insights from leading scientists, ethicists, and thought leaders, Decoding Life offers a balanced and nuanced perspective on topics ranging from reproductive technologies and gene editing to bio banking and synthetic biology. This section seeks to promote informed dialogue, critical reflection, and responsible decision-making in our fast evolving world by exploring the ethical, social, and philosophical aspects of genetic engineering. This journey will challenge our assumptions, provoke our conscience, and contemplate the future of humanity.

Ancient civilisations held a variety of diverse and conflicting beliefs regarding the origins of life and the development of distinct species. Anaximander, a philosopher who lived from 610 to 546 BCE, theorised that land animals were derived from sea animals.[67]

Aristotle, the renowned philosopher and polymath of ancient Greece, who lived from 384 BCE to 322 BCE, believed that life emerged spontaneously, a concept common among scholars of his era. He held the view that flies came from decaying matter, mice were created from hay, and crocodiles came from decaying wood in water. However, Aristotle, made one of the first attempts to explain life's diversity. He introduced the concept of "Scala Nature" or "Great Chain of Being." This concept organised all living creatures in a hierarchy. Plants were at the bottom, while humans were at the top, symbolising the highest form of life. His hypothesis was a significant step forward in comprehending life's complexity.[68,69]

Fig 2: *Aristotle 384BCE- 322 BCE*

Abiogenesis is the leading theory about the origin of life on Earth. It suggests that life emerged in a warm, ancient pond filled with key chemical elements for life- carbon, hydrogen, nitrogen, oxygen, phosphorus, sulphur, etc. It is postulated that lightning strikes provided the energy needed to spark the formation of complex molecules like RNA and DNA. DNA is primarily important because of its unique ability to replicate/ reproduce.[206]

"The aim of science is not to open the door to infinite wisdom, but to set a limit to infinite error."

Bertolt Brecht

In 1871, Charles Darwin famously wrote to his friend Joseph Dalton Hooker, proposing that life could have started in a "warm little pond" with various chemicals like ammonia and phosphoric salts. He speculated that light, heat, and electric sparks might have acted on these compounds to create proteins, which later evolved into more complex forms of life. Alexander Oparin and J.B.S. Haldane were other proponents of this theory. Oparin, a Russian scientist, is considered the 20th-century Darwin. This is the prevalent theory accepted by the scientific community.[56,95,192]

Fig 3: Charles Robert Darwin **(1809 –1882);** Alfred Russel Wallace **(1823-1913)**

Pioneering work by Stanley Miller in 1952, under the supervision of Nobel laureate Harold Urey at the University of Chicago, produced building blocks of life from non- living substances. The discovery of DNA (1950) and RNA, along with their catalytic activity and self-assembling lipids, forming membranes, suggests the emergence of life through natural selection.[176, 262]

DNA acts like a recipe book, outlining the instructions for making proteins, which in turn define the unique features of every living being, from single-celled organisms to the exalted humans. All life shares this same genetic code, suggesting a common ancestor known as the 'Last Universal Common Ancestor' (LUCA). While some viruses use RNA instead of DNA, they still need a DNA-based host to reproduce.

Over time, life evolved, and organisms with multiple cells started appearing approximately 1.6 to 2 billion years ago. Around 2 billion years ago, a significant evolutionary change occurred when living organisms started using oxygen to produce energy. Cyanobacteria, which released oxygen into the atmosphere made this shift possible. Using oxygen for energy production (aerobic metabolism) is much more efficient and helped to speed up the pace of evolution. These changes eventually culminated in the 'Cambrian Explosion' about 540 million years ago—a period when many new types of life forms emerged and diversified. This event was one of the most significant evolutionary milestones in

Earth's history, leading to the development of nearly all the major body structures (phyla) we see today.[52, 158, 234]

From their humble beginnings, single-celled organisms multiplied and experienced genetic mutations. A mutation is a change in the DNA of a living organism. It can happen by accident when cells copy their DNA or because of outside factors like radiation. Mutations can change how a gene works, sometimes causing differences in traits, like eye colour or even leading to diseases. Most mutations are harmless, but some can be helpful or harmful, depending on how they affect the organism. Some mutations help in the adaptation to the surrounding environment. Individuals with beneficial mutations thrive, out-competing others and leaving more offspring, ensuring the propagation of those beneficial traits. Accumulation of multiple mutations and geographical separation (allopatric speciation) over a long period leads to new species and the vast biodiversity we see today. Now, Earth is home to about 9 to 10 million species, spanning plants and animals.[6, 54,279]

Asexual and Sexual Reproduction

Asexual reproduction, common in single-celled organisms, plants, and fungi, enables rapid propagation without fertilization. While it allows fast regeneration, it lacks genetic diversity, limiting long-term adaptability. Binary fission, budding, vegetative reproduction, spore formation, and fragmentation are key forms. Parthenogenesis, where an unfertilized egg develops into a new individual, occurs in species like whiptail lizards, Komodo dragons, and certain insects.[58,269,270]

Sexual reproduction emerged 1.2 billion years ago, enhancing genetic diversity and adaptability. Most eukaryotes reproduce sexually, with each parent contributing half of the offspring's genes. This fosters genetic variability, activates suppressed genes, counters harmful mutations, and promotes complexity. Some fungi and bacteria have multiple sexes, further diversifying evolution.[58,81]

The reproductive system evolved from binary fission through transformation, transduction, and conjugation to sexual reproduction. Approximately 99.9% of eukaryotes reproduce sexually, a process where each parent contributes only half of the offspring's genes, fostering genetic variability and adaptive advantages. Suppressed genes may activate over generations, recombination counteracts harmful mutations, and sexual reproduction encourages complexity, improving survival.[81]

Survival involves both adaptation and overcoming challenges. Evolution favors progression rather than mere existence, ensuring species develop resilience.This extends to metabolism, reproduction, form, and function. Sexual reproduction introduced partner selection, advancing species fitness over generations. While evolution is non-teleological, driven by natural selection rather than a set goal, it has led to the refinement of the nervous system and brain.

Carl Sagan noted in *The Dragons of Eden* that brain-stored information eventually surpassed genetic information in evolutionary complexity, from reptiles onward.[54,224]. (Interestingly, in the story of genesis in the Bible, the fruit of knowledge is given to 'Eve' by a reptile) Single-celled organisms do not age but multiply through binary fission, dying only from injury or starvation. Conjugation allowed genetic mixing, leading to multicellular organisms with specialized functions, accepting death as part of aging. Vegetative reproduction in plants mirrors this phenomenon. Sexual differentiation and reproduction accelerated evolution, enabling gene mixing and selection.[99, 174.]

Chapter 3

EVOLUTION

From Lamarck to Darwin

"Nothing in biology makes sense except in the light of evolution."

Theodosius Dobzhansky

While ancient philosophers speculated about evolution, Jean-Baptiste Lamarck (1744–1829) was the first to propose a comprehensive theory. He coined the term "biology" and studied invertebrates, classifying crustaceans, arachnids, and molluscs. Lamarck suggested that traits acquired during an organism's life-like a giraffe stretching for leaves-could be inherited. Though dismissed in his time, his ideas found echoes in modern epigenetics.

Lamarck believed organisms had an innate drive toward complexity, climbing a "ladder of life." While modern science favours adaptation over perfection, debates on evolution's purpose continue. His life was marked by financial struggles, and he died in poverty. However, his work remains foundational in evolutionary thought.

The first convincing theory of evolution came from Charles Darwin and Alfred Russel Wallace. Darwin, after years of research-including his voyage on the HMS Beagle-hesitated to publish his theory. Meanwhile, Wallace, a self-taught naturalist, independently conceived natural selection in 1858 while in the Malay Archipelago. His letter to Darwin, now known as the "Ternate Letter," shocked Darwin, as it mirrored his own conclusions.

To ensure fair recognition, Darwin's peers arranged a joint presentation at the Linnean Society in 1858, but it went largely unnoticed. The following year, Darwin's *On the Origin of Species* was published, selling out on the first day. More comprehensive than the joint presentation, it became the cornerstone of evolutionary biology.

Charles Darwin wrote in The Origin of Species: "It is interesting to contemplate an entangled bank, clothed with many plants of many kinds, with birds singing on the bushes, with various insects flitting about, and with worms crawling through the damp earth, and to reflect that these elaborately constructed forms, so different from each other, and dependent on each other in so complex a manner, have all been produced by laws acting around us."

"Nature is an endless combination and repetition of a very few laws. She hums the old well-known air through innumerable variations."

Ralph Waldo Emerson

The Evolutionary Journey to Humans

Around 400 million years ago, plants developed seeds, and insects evolved to pollinate them, creating a symbiotic relationship that shaped ecosystems. In the oceans, fish with fins gradually developed limbs, allowing them to move on land. By 360 million years ago, the first amphibians emerged, bridging water and land life.

***Fig 4**: Vertebrate evolution*

Primates evolved traits like opposable thumbs and stereoscopic vision, setting the stage for human evolution. Around 5–6 million years ago, our ancestors split from chimpanzees, with whom we still share 98.4% of our DNA. Our lineage diverged from gorillas even earlier, 8–10 million years ago, making them our "first cousins" in evolutionary history.

Despite millions of years of evolution, the deep genetic connections between all living species reveal the intricate web of life. As Darwin poetically described, nature is an "entangled bank" of interconnected forms, shaped by the forces of evolution.

Humans by Natural Selection

"The story of evolution unfolds with increasing levels of complexity, and the beauty of it is that each level brings new laws and new realities."

Edward O. Wilson

A common misconception is that humans evolved directly from modern primates. In reality, we share a common ancestor with them. Around 4 million years ago, as Africa's rainforests gave way to open savannahs, early human ancestors adapted by walking upright. This shift began with *Ardipithecus*, followed by *Australopithecus afarensis*—famously known as "Lucy," the mother of humankind. Over time, brain size increased, and Homo sapiens eventually emerged, distinguished by intelligence and complex social behaviours. [93]

While we pride ourselves on intellect, philosophers like Nietzsche and Camus question whether intelligence has truly led to happiness. Unlike other species, we grapple with meaning, mortality, and existential anxiety. Could intelligence be an evolutionary experiment rather than an ultimate goal? Other life forms might have evolved simpler, more harmonious ways of being, avoiding the burden of self-awareness. Perhaps other phyla have avoided the 'trap' of intelligence by evolving simpler. [30,186.] The uniqueness of human intelligence could be seen as a dual-edged sword, an extraordinary development that enables creativity, understanding, and capabilities to seek meaning and purpose, but also alienation and existential angst

Fig 5: *The tree of evolution of primates from 60 million years ago to the present*

The Evolutionary Timeline

- **30 mya** – New World monkeys (Americas)
- **25 mya** – Old World monkeys (Africa, Europe, Asia)
- **10 mya** – Gorillas emerge
- **6 mya** – First hominids appear
- **3.6 mya** – *Lucy* (*Australopithecus afarensis*)
- **2.3 mya** – *Homo habilis* (early tool user)
- **1.8 mya** – *Homo erectus* (spread beyond Africa)
- **250 tya** – *Homo* Neanderthalensis
- **100 tya** – *Homo sapiens*

Fig 6: *Australopithecus afarensis (Lucy and Family)*

Evolution of Intelligence, Function and Beauty

"A living organism always does something unexpected, because it is alive."

Ludwig von Bertalanffy

Human evolution has been shaped by environmental and climatic changes, and today, in the Anthropocene, we have an unprecedented impact on the planet.

Early ancestors like *Australopithecus afarensis* (*Lucy*) had small bodies, long arms, and short legs—features suited for both tree-dwelling and upright walking. Their plant-based diet required larger stomachs, showing how adaptation was key to survival.[126,137,156]

Lucy's discovery in 1974 by Donald Johanson's team was ground-breaking. Found in Ethiopia, she was one of the most complete early human fossils. Her name came from The Beatles' song *Lucy in the Sky with Diamonds*, which played at the excavation site. [127]

Brain Growth and the Future

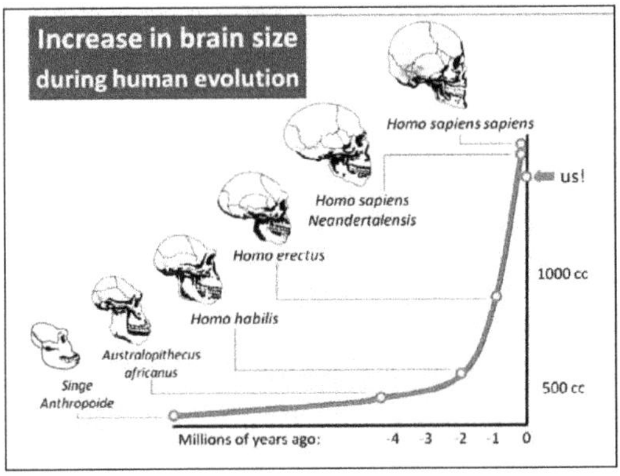

***Fig 7:** The volume of the brain has increased tremendously during evolution The increase in size was steep from 2 million years ago, from our ancestor Homo habilis*

(For millions of years, brain size grew slowly. But around 2 million years ago, with *Homo habilis* (the "handy man"), brain volume increased rapidly. . Later species, like *Homo heidelbergensis*, had even larger brains—up to 1600cc, bigger than modern humans. Yet, intelligence isn't just about size—culture, tools, and language have been just as vital. Today, our brains average around 1450cc, but technology may change our trajectory. With AI and brain-computer interfaces, could human brains shrink as we offload cognitive tasks to machines? [63, 200.] The future may bring enhancements like memory implants and direct internet access, raising ethical questions we'll explore later.

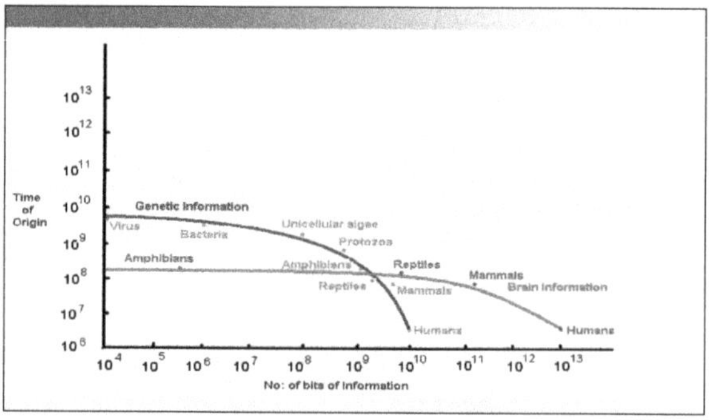

Fig 8: *Comparison of bits of information in the genes and in the brain accords an interesting observation. For amphibians the information in the chromosomes is much more than the information in the brain. But it has reversed in mammals and maximum in humans. In the graph we find that the reversal started in a reptile. We can observe that from then on, the focus of evolutionary thrust was more on the brain than on the body.*[224]

(' Idea taken and modified from Carl Sagan's book 'Dragons of Eden ')

***

(' As humans climbed the ladder of evolution
Jaws lost many a function
Size suffered a gradual reduction
Brain enlarged for better cognition
The seat of special sensory reception
Face moors the fine human expression
In the prolonged evolutionary bargain
Unique smile was the charming gain.
AI, soon may make a digital brain
Humans may suffer an existential strain ')

The Role of Beauty in Evolution

"Evolution is cleverer than you are.

Leslie Orgel

"If everything were cast in the same mould, there would be no such thing as beauty."

Charles Darwin

Sexual selection is a powerful force in evolution. As Darwin observed, organisms compete not just for survival but also for mates. This struggle takes two main forms: physical dominance and the ability to attract a partner through charm and beauty.

To impress females, some, like bower birds, even build intricate nests, suggesting that the appreciation of beauty predates human existence. [9, 81, 138, 223]

Fig 9: *Bower bird nest (from National geography site)*

In humans, beauty has deep evolutionary and social roots. Traits like facial symmetry, clear skin, and proportional features have been linked to genetic fitness. Since childhood, attractive individuals often receive better treatment, reinforcing societal standards of beauty.

The Golden Ratio: Nature's Formula for Beauty

"Geometry has two great treasures: one is the Theorem of Pythagoras; the other, the division of a line into extreme and mean ratio. The first we may compare to a measure of gold: the second we may name a precious jewel."

Johannes Kepler

The **Golden Ratio** (approximately 1.618) appears throughout nature and art, shaping the proportions of flowers, seashells, and even the human body. Artists like

Leonardo da Vinci and Michelangelo used it to create visually harmonious works.

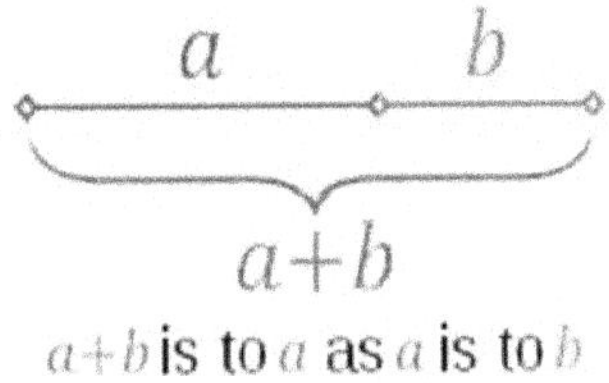

Fig .10: *A+B=C; A/B = C/A = 1.618*
A/C = B/A = 0.618.

The Fibonacci Sequence

1,1,2,3,5,8,13,21,34,55,89,144,233,377...

1+1=2	**13+21=34**
1+2=3	**21+34=55**
2+3=5	**34+55=89**
3+5=8	**55+89=144**
5+8=13	**89+144=233**
8+13=21	**144+233=377**

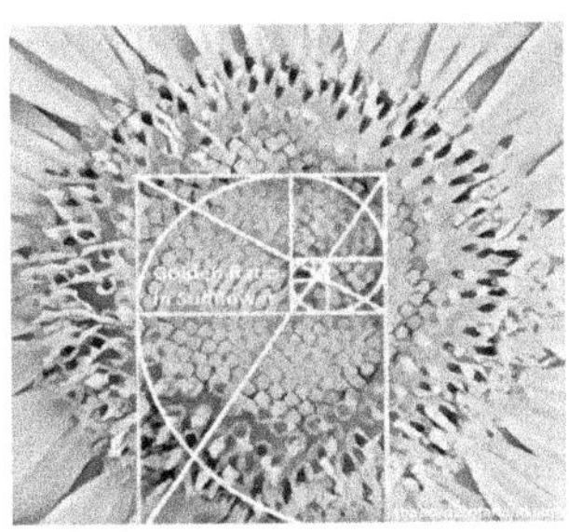

Fig11: *Sunflower depicts golden propotion in the arrangement of petals.*

This principle is also reflected in the **Fibonacci sequence** (0,1,1,2,3,5,8...), where each number is the sum of the previous two. Many natural patterns, from leaf arrangements to galaxy spirals, align with this ratio, suggesting that beauty and function are intertwined[110, 157, 163].

Beauty: A Universal, Yet Cultural Concept

While mathematical patterns define balance and proportion, human perception of beauty is influenced by culture. What is considered attractive varies across societies and eras.

Fig 12: *Michel Angelo's famous painting in Sistine Chapel. 'Seduction and expulsion of Adam and Eve.' He has used the golden proportion in this painting.*

Fig 13. *Venus de Milo at the Louvre has been described as a "classical vision of beauty."*

Plato believed beauty lies in harmony and proportion, a principle that nature seems to follow. But beyond numbers, the emotional and symbolic impact of beauty is what truly captivates us.

Mate Selection and Beauty

"The success of an animal in the struggle for existence is not measured only by the number of offspring which it produces and rears, but also by the probable success of these offspring."

Ronald Fisher

Physical attraction is shaped by social, cultural, and subjective preferences, with intelligence and social status often linked to genetic fitness. Freud associated beauty with sexual excitement, while evolutionary psychologists view it as an indicator of reproductive success.[79,131]

Body proportions matter-ideally, 45% of the breast volume should be above the nipple and 55% below. The waist-to-hip ratio of 0.7 is considered most attractive, aligning with the golden ratio (0.618). Evolution may be gradually shaping human proportions toward this ideal. [73 89,162,209,223]

Men are attracted to youthful features—full breasts, symmetrical faces, and low waist-to-hip ratios—while women prefer taller men with broad shoulders and masculine features. Beauty, however, does not always correlate with happiness. Facial symmetry and proportion play a crucial role in attraction, with balance and harmony appealing to human perception. [35, 145, 181]

Attractive people are often assumed to be more capable and successful. Camouflaging through clothing enhances perceived beauty—women use padding and high heels, while men opt for thick-soled shoes. Large breasts and buttocks may have evolved as secondary sexual characteristics, compensating for the hidden female genitalia due to bipedalism. Human breasts remain enlarged even when not lactating, unlike in other mammals, possibly serving as a visual cue for fertility. [8,22, 94,]

Childlike features such as large eyes and smooth skin are universally attractive. Plastic surgery often mimics these traits to enhance appeal. Studies show that babies prefer pretty faces, indicating an innate attraction to symmetry and proportion.[129]

Body odor also influences attraction, with studies showing women prefer the scent of physically attractive men. Napoleon's letter to Josephine, "I will return tomorrow—don't wash," underscores the primal nature of scent in attraction.

Sexual selection drives partner choice. Traits indicating health, vitality, and intelligence become attractive by default. Darwin noted that appearance influences mate selection, and studies confirm that attractive women often marry high-status men. Power and resources, like beauty, play key roles in mate preference. 110, 139, 145, 154,

"The forms of nature are endlessly beautiful, and their beauty derives from their function."

D'Arcy Wentworth Thompson

Evolution favours traits that signal fitness. This is evident in animals—modern horses and elephants possess both grace and functionality. Humans, too, have evolved toward greater refinement, as seen in Leonardo da Vinci's Vitruvian Man, which illustrates divine proportions. Proper proportion enhances efficiency, aesthetics, and survival.

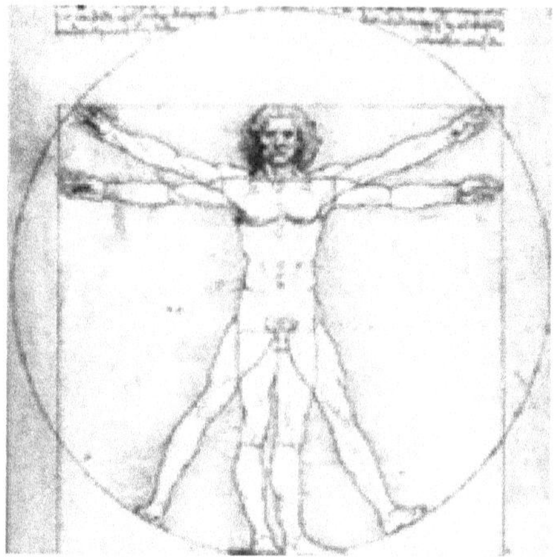

Fig 14: *Vitruvian man.*

A ***B***

Fig 15: Evolutionary ladder of horse (A) and elephant (B) *(Showing the trend towards beauty and fitness.)*

Beauty standards vary across cultures and time, yet certain traits remain universally appealing. Shakespeare stated "*Beauty lies in the eyes of the beholder,*" while Keats glorified its lasting joy."*A thing of beauty is a joy for ever*" True beauty radiates vitality and happiness, influencing perception and social success. [136,229,230]

Face, Brain, and Beauty

"The human brain is the most complex device in the known universe, yet it is an imperfect one."

Dean Buonomano

Physical and intellectual evolution shaped hominids, prioritizing brain expansion over brute strength. The human brain is three to four times larger than that of apes, extending lifespan and enhancing cognition. Larger brains required evolutionary trade-offs, shifting emphasis from raw physicality to intelligence. Hominins developed tools as "*extra limbs,*" reducing the need for physical adaptations while increasing cognitive demands. Recent studies indicate brain size has slightly decreased over the past 20,000 years. Efficiency, not sheer size, is key—social cooperation and technological advancements may compensate for smaller brains.

Looking ahead, human brains could integrate with technology, enhancing memory and intelligence. Future evolution might favour neotenous traits—larger heads, bigger eyes, and smaller jaws—features already perceived as attractive. Beauty and intelligence are intertwined. Over time, selecting the "prettiest and brightest" contributes to evolution, ensuring survival and progress.

"The peacock's tail is the most elegant of proofs that beauty is not just a human concern—it is woven into the evolutionary fabric of life itself."

Richard Prum

Nancy Etcoff, in *Survival of the Prettiest*, notes that in nature, beauty signals reproduction. Animals display vivid colors at sexual maturity—peacocks unfurl psychedelic feathers, and flowers attract pollinators with colors and scents. White flowers like jasmine emit fragrance for nocturnal insects, while orchids mimic female wasps to lure males into aiding pollination. Some flowers even reward pollinators with nectar.

"The good, of course, is always beautiful, and the beautiful never lacks proportion."

Plato

The human brain may continue evolving through technology, with implants enhancing memory, accelerating learning, or enabling internet connectivity. While promising, these innovations raise ethical concerns and risks that require careful consideration. We will explore these issues later.

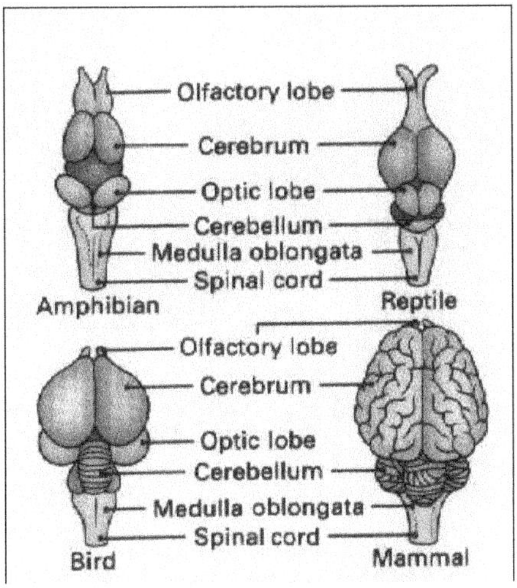

Fig 16: *Cerebrum controls intelligence. Comparison of brains from Amphibian to Mamma*

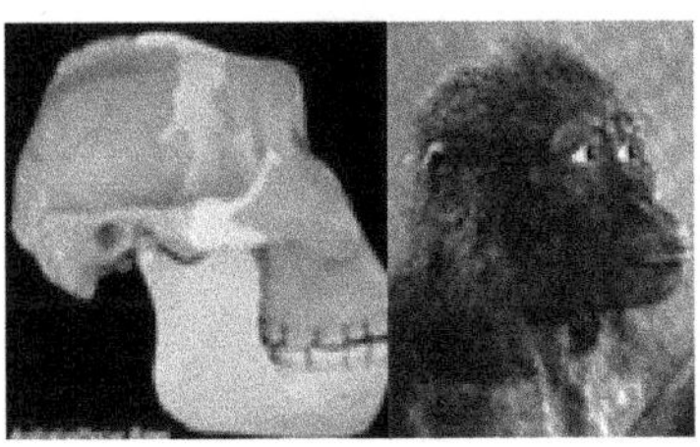

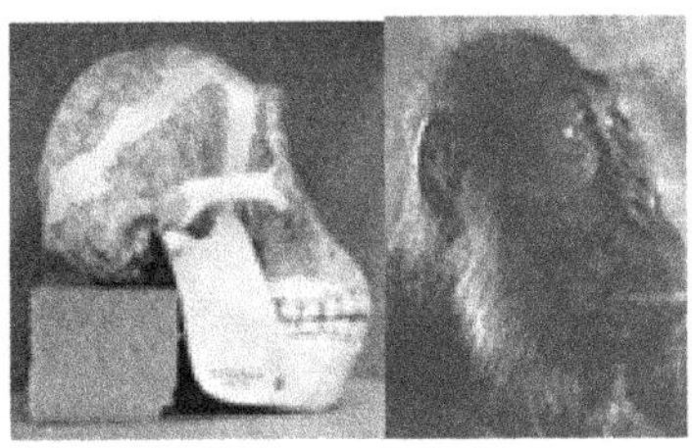

Australopithecus afarensis ***Aboisei***

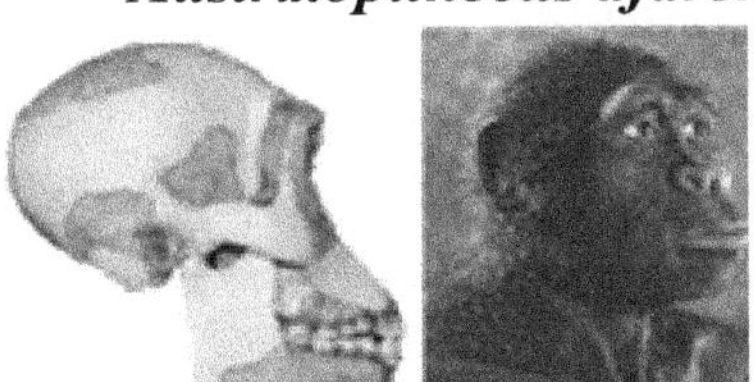

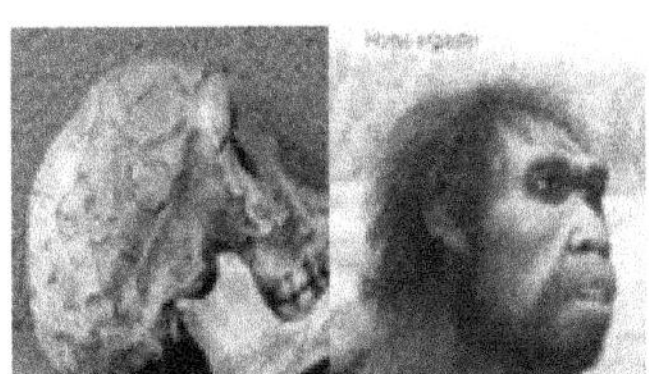

Homo habills **H. ergaster**

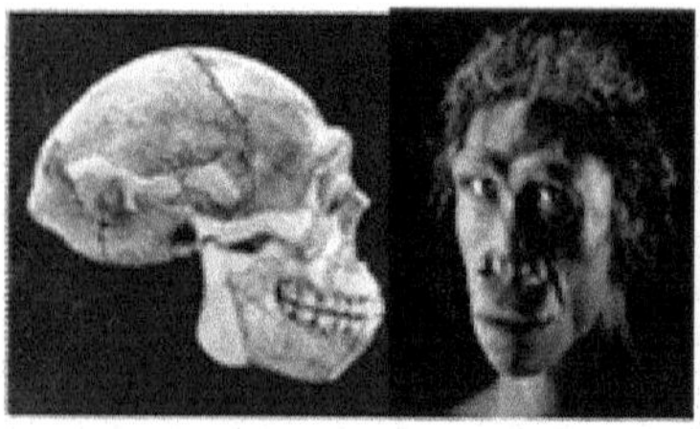

H. erectus

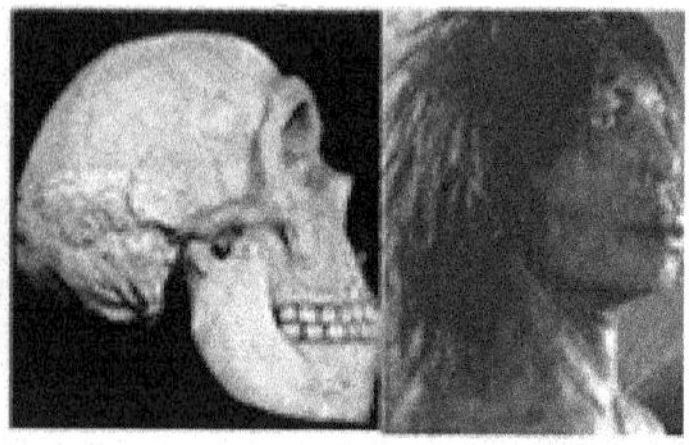

H. neanderthalensis

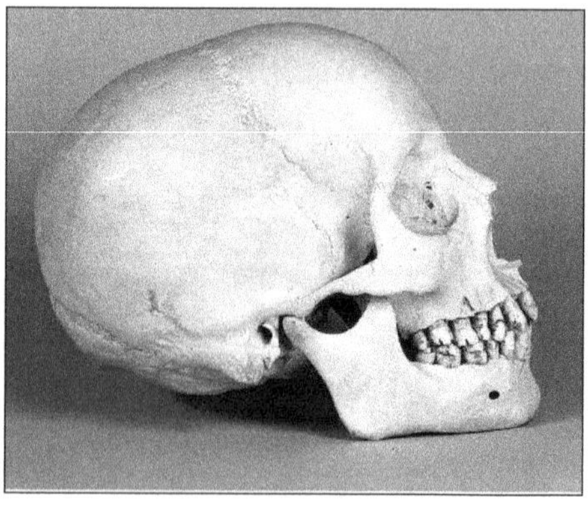

Homo sapiens.

Fig 17: *Our Ancestors and Homo sapiens (The transformation of the human face and skull from Australopithecus afarensis (Lucy) to Homo sapiens is shown in this picture)*

Skull Face proportion: When we discuss the proportions for beauty it is imperative that we consider Phi (f) the golden proportion which is 1:1.618.

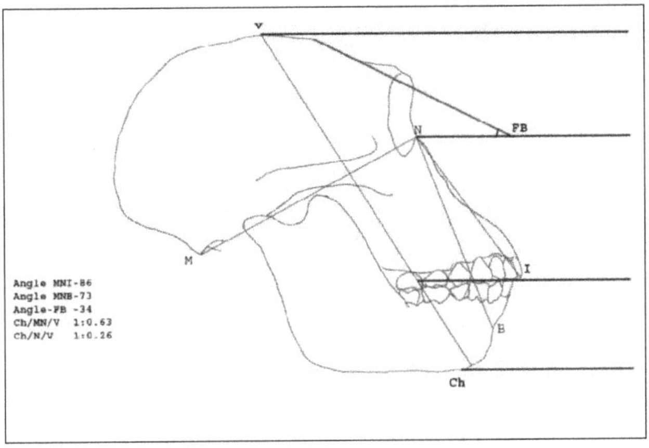

A

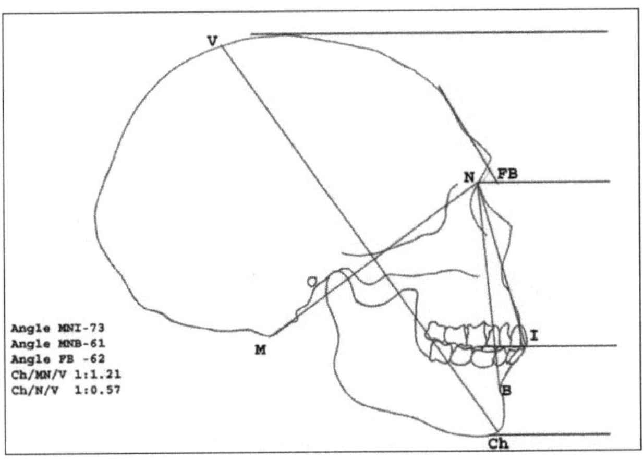

B

***Fig18:** Cephalometric analysis of Australopithecus (early human ancestor) (A) and the modern human (Homo sapiens)(B).*

Cephalometric analysis to assess and compare the following parameters of different species of the Hominidae family. 1) Maxillary projection.(angle MNI) 2) Mandibular projection.(angle MNB). 3) Frontal bossing. (angle FB). 4) Cranium- facial proportion in two planes- frontal and lateral.

Species	Years MYA	Angle	Angle	Angle	Ch/MN/V	Ch/N/V	Brain Vol.(cc)
		MNI	MNB	FB			
A.africanus	3	86	73	34	0.63	0.26	420-500
A bosei	2.5	80	70	26	0.7	0.27	480
H.habilis	2	78	69	51	0.92	0.5	650
H.ergaster	1.5	76	65	43	1.06	0.5	850
H.erectus	1	75	65	46	1.06	0.5	800-1000
H.neanderthalensis	0.5	74		50	1.05	0.51	1400-1600
H.sapiens	0	73	61	62	1.21	0.57	1040-1450

Table 1: The prominence of the upper jaw (Angle MNI) has reduced from 86 degrees to 73 degrees); the same way the prominence of lower jaw (Angle MNB) has reduced from 73 degrees to 61 degrees during the last 3 million years. But the prominence of the frontal bone which is the seat of frontal lobe of the brain, the seat of intelligence, has increased from 34 degrees to 62 degrees (FB), This indicates the increase of intelligence and the enlargement of the brain. C/MN/V is the proportion between the face and the cranium on profile view. CH/N/V is the proportion between the face and the forehead from the front view.

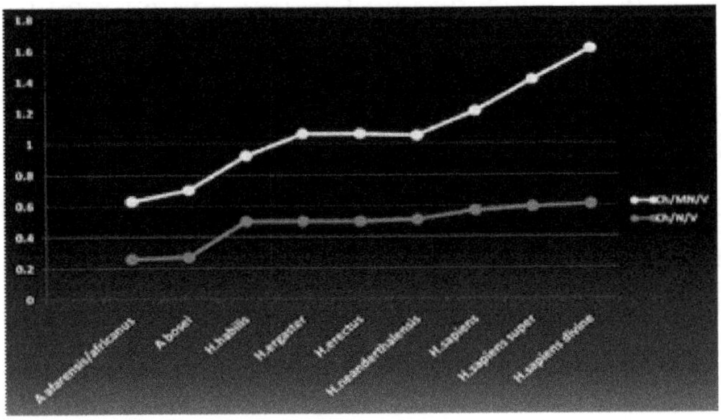

Fig 19: *With the increase in the brain volume, the proportion between the cranium and the face is becoming nearer to the golden proportion. The extrapolated graph shows that to reach the golden proportion, it may take one million years and two more species by natural means. (If there is no hindrance by genetic engineering, nano-technology and AI.)*

Fig 20*: Future human: When the proportion reaches golden. The face may become oval with large eyes, large forehead having a large frontal lobe of the brain.*

Thinkers like Nietzsche and Plotinus envisioned humans evolving toward intelligence and beauty, but predicting future human evolution remains highly speculative. Some theories suggest that in a million years, new human species may emerge with neotenous (childlike) features—larger heads, bigger eyes, and smaller jaws—aligned with "golden" proportions. This trend, called neoteny, preserves juvenile traits into adulthood.

Interestingly, human brains have shrunk over the past 20,000 years, yet intelligence has increased. (?). The article *If Modern Humans Are So Smart, Why Are Our Brains Shrinking?* suggests this may be due to improved cooperation, collective intelligence, and energy efficiency. Some believe brain reduction signals self-domestication, much like in tamed animals.[133]

Chapter 4

NATURE'S BIOTECHNOLOGY

"Our understanding of evolution has changed radically in the last century. What once seemed a blind, mechanical process now appears more intricate and self-organizing than we ever imagined."

James Shapiro

With today's rapid technological and societal advancements, human evolution may soon change in ways that go beyond natural processes. The future of humans might be determined by unknown factors. Historically, evolution was slow and driven by natural selection. But now, technologies like genetic engineering, nanotechnology, and medical break throughs could speed up and shape human evolution in new and unexpected ways.

These advancements may allow us to enhance our physical and intellectual capabilities. While natural selection still has a role. Humans could soon manipulate their own genes, changing the course of evolution. For most of human history, we adapted to our environment to survive. Changes in our biology, behaviour, and culture were responses to natural challenges.

For billions of years, life has evolved through processes like 'descent with modification, adaptation, natural selection, and the survival of the fittest.' These mechanisms have shaped the diversity of life on Earth. At the heart of this evolution lies *deoxyribonucleic acid* (DNA)—a remarkable molecule that can replicate and is the fundamental building block of life. Understanding DNA provides insight into the resilience, diversity, and interconnectedness of living organisms, highlighting the complexity and beauty of the natural world.

Genes, DNA and chromosomes

"The genome is a book that wrote itself, continuously adding, deleting, and rewriting over four billion years."

Matt Ridley

Genes are the basic units of heredity that pass from parent to offspring. DNA sequences make up genes, and chromosomes in the cell nuclei arrange them. Our traits—whether physical or behavioural—are determined by the genes to a large extend, embedded in these chromosomes.

DNA, the core component of chromosomes, has a double-helix structure resembling a twisted ladder. It is made of four nucleotides: Adenine (A), Thymine (T), Guanine (G), and Cytosine (C). These nucleotides pair up in specific ways: A pairs with T, and G pairs with C. A codon, consisting of three consecutive nucleotides, can be one of 64 possibilities (4x4x4). However, only 61 codons code for amino acids—the building blocks of proteins—while the other three serve as stop signals in the genetic code. Since there are only 20 amino acids, multiple codons can produce the same amino acid.

A series of codons form a gene, and non-coding segments of DNA separate genes. In a metaphorical sense, nucleotides are like letters, codons are words (representing amino acids), genes are sentences (proteins), and chromosomes are the books that define our characteristics.

Structure of Chromosome and DNA

"We are all, in a sense, the children of DNA. It is our real, eternal ancestor."

Richard Dawkins

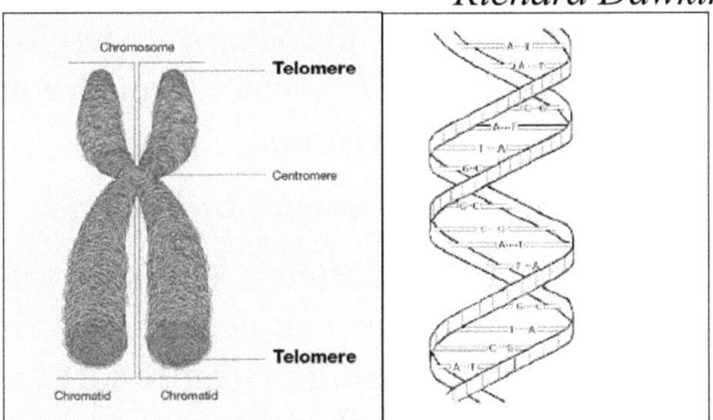

Fig 21: *Genes are the basic unit of heredity transferred from parent to off spring.*

Genes are made up of sequences of DNA and are arranged, one after another, at specific locations on chromosomes in the nucleus of cells.

In sexually reproducing organisms, chromosomes come in pairs—one from each parent. Humans have 23 pairs of chromosomes, with each parent contributing one chromosome to each pair. Together, these chromosomes contain 20,000 to 25,000 genes, responsible for our physical and mental traits.

Interestingly, a single gene can produce multiple proteins through a process called gene splicing. This process significantly increases the complexity of proteins that a few genes can produce.[66]

Life on Earth began with a protein and then single-celled organisms that reproduce by binary division-a process in which one cell splits into two identical cells, passing on identical genetic material to both daughter cells. Hence, there is no genetic diversity. In contrast, sexual reproduction introduces variety by producing specialised cells, called gametes through a process called meiosis (reduction division)..

Horizontal Gene Transfer in Bacteria

However, single-celled organisms are masters of survival and can exchange genetic material with each other. Bacteria, though simple organisms, have fascinating ways to exchange genetic material. They can engage in horizontal gene transfer through three principal methods. This allows them to share traits and a dapt quickly to changing environments.

Transformation: When bacteria die, their DNA can break apart and float in the environment. Other bacteria can take up these DNA fragments and incorporate them into their own genomes.

Transduction: A virus that infects bacteria, called a bacteriophage, can transfer DNA from one bacterium to another.

Conjugation: Two bacteria can connect through a structure called 'pilus', allowing them to exchange plasmids—small, circular DNA molecules. This process is often referred to as 'bacterial sex' and is considered a precursor to sexual reproduction.[190, 255]

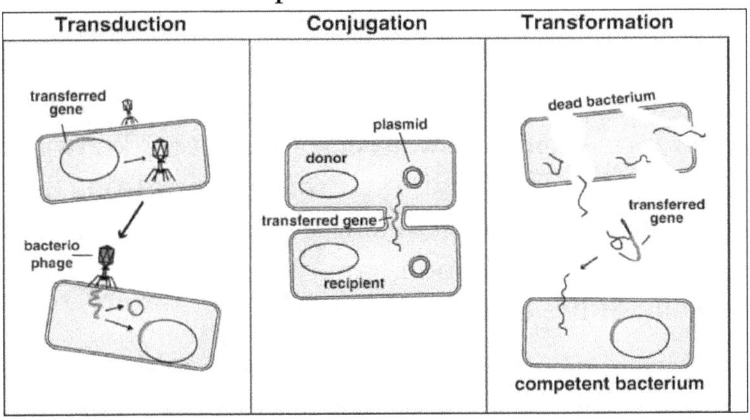

Fig 22; Horizontal gene transfer

As the evolutionary ladder moves up to reptiles, birds, and mammals, fertilisation occurs internally. This adaptation is useful for land-dwelling animals, as it protects the gametes from drying out or being eaten by predators. In mammals, including humans, fertilisation happens inside the female's reproductive tract, where sperms are deposited during mating. This internal process allows for greater control over reproduction and protection of the developing embryo, and often leads to internal gestation (pregnancy) and parental care.

The Role of Genetic Diversity in Evolution

Sexual reproduction introduces genetic variation through meiosis, mutations, and environmental influences. This variation helps species adapt to changing environments and increases their chances of survival. The fitter individuals have better survival chances, get better partners, produce more offspring and live longer. They pass on their traits to the next generation, a key principle in evolution. Over time, these genetic changes can lead to the formation of new species, a process known as ' speciation.' This occurs through mutations, gene variations, environmental shifts, and migrations. These processes drive evolution, contributing to the incredible diversity of life on Earth[58, 81, 170]

(Darwin's Finches: On the Galápagos Islands, Darwin observed finches with varying beak shapes, adapted to different food sources. Darwin's finches became a famous example of adaptive variation where species evolve from a common ancestor to fill different ecological niches, helping cement the theory of evolution by natural selection.)

Learning from Nature: Selective Breeding and Its Consequences

For thousands of years, humans have unknowingly "hacked" nature through 'Selective Breeding'—choosing plants and animals with desirable traits to reproduce, shaping the species we rely on today. This method has been used to develop various breeds and varieties within the same species.

One well-known example is the domestic dog, which descended from the wild grey wolf (*Canis lupus*). By selecting wolves with specific traits, humans have bred and reared a wide variety of dogs, ranging from tiny poodles to powerful bulldogs. Despite their differences in size, behaviour, and appearance, all dog breeds belong to the same species, *Canis lupus familiaris*. Selective breeding has allowed humans to create animals suited for specific purposes-such as herding, hunting or companionship-demonstrating our ability to influence nature, often unintentionally.

Fig 23: *From Grey Wolf to Modern Dog Breeds*

Selective breeding also transformed plants. The Brassica plant, for example, has given rise to an array of vegetables including kohlrabi, cauliflower, broccoli, and kale. These foods are the result of careful selection over generations, shaping the original plant into the nutritious vegetables we eat today.

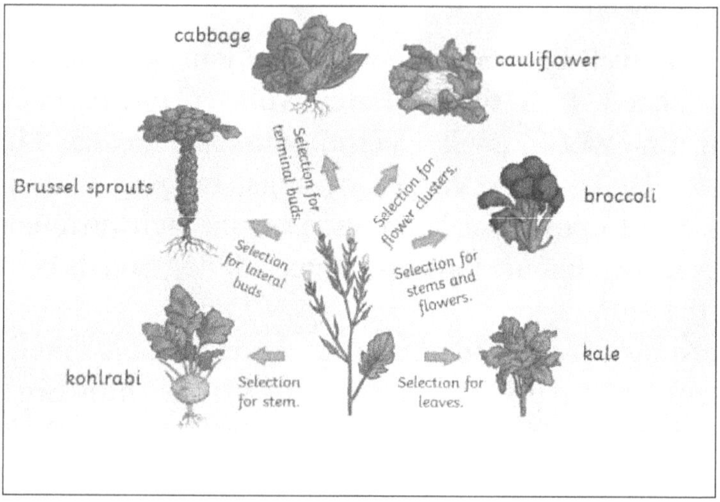

Fig 24: *From Brassica to Kale, Cauliflower, Broccoli, Etc.*

The Dark Side of Selective Breeding: Eugenics

While people have used selective breeding for beneficial purposes, they have also misused it in harmful ways. In the late 19th and early 20th centuries, Sir Francis Galton, a cousin of Charles Darwin, advocated for eugenics-a movement aimed at improving the genetic quality of humans by promoting desirable traits and discouraging undesirable ones. Eugenicists believed that controlled breeding could eliminate social problems and enhance human characteristics.[196]

However, this idea was deeply flawed. The most infamous misuse of eugenics occurred during Adolf Hitler's Nazi regime, where the foible and faulty concept of a 'superior Aryan race' was used for horrific policies of eccentric racial purity. Millions of Jews, Romani people, disabled individuals, and others deemed "undesirable" were murdered during the Holocaust. This tragic chapter in history serves as a warning about the dangers of pseudoscientific beliefs and the ethical misuse of genetics. The lasting effects of the atrocities committed under the guise of eugenics emphasise the necessity of ethical considerations and scientific precision in dealing with genetics and any endeavours to alter human characteristics.[271, 208] (Adolf Hitler, the dictator of Nazi Germany, died by suicide on April 30, 1945, in his underground bunker in Berlin. Facing imminent defeat as Allied forces closed in on the city during the final days of World War II, Hitler took his own life alongside his long-time partner, Eva Braun, whom he had married the day before. Their deaths marked the collapse of the Nazi regime, leading to Germany's surrender on May 7, 1945, and the end of the war in Europe. Hitler's death remains one of the most significant events in 20th-century history.)[14]

Ethnic conflict remains a significant issue in many parts

of the world even today, fuelled by historical grievances, political manipulation, and social inequalities. From the persecution of Rohingya Muslims in Myanmar to the ethnic violence in Ethiopia's Tigray region, these conflicts often arise from deep-seated divisions between groups based on identity, culture or religion. The present war between Israel and Palestine also traces its roots to ethnic factors. Despite international efforts for peace, ethnic conflicts continue to challenge global stability.[108]

Chapter 5

GENETIC ENGINEERING

"We can now manipulate life at its most fundamental level. The question is no longer 'can we?' but 'should we?'"

Jennifer Doudna

In the ancient world, before the development of modern scientific methods, people often attributed the origins of life to mystical or spontaneous forces. For example, the belief was widespread that flies emerged from decaying fruit or that crocodiles were born from submerged logs. These notions remained unchallenged until scientific exploration provided more concrete explanations.

Unveiling the Secret of Life

Antonie van Leeuwenhoek's invention of the microscope brought a breakthrough in biology. In 1677, Leeuwenhoek made an astonishing discovery when he examined his semen under the microscope. He observed tiny, tadpole-like creatures. (which we now know as sperm cells). He called them "*animalcules*" (little animals), believing they each carried a tiny human, a theory known as preformationism. Renowned as "The Father of Microbiology," Antonie van Leeuwenhoek (1632-1723) was a Dutch tradesperson and scientist who made significant advancements in the field.

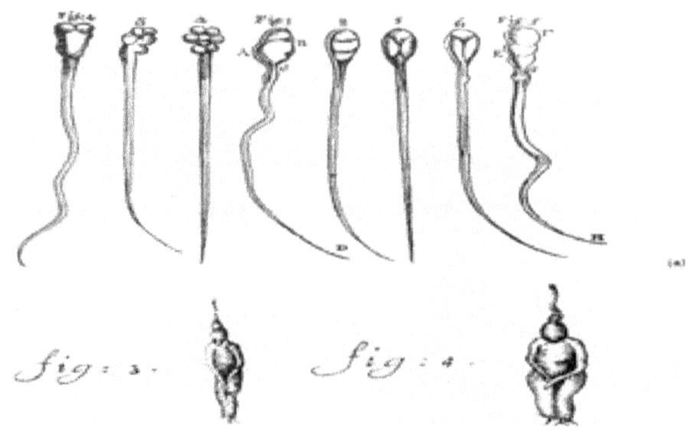

Fig 25: *The 17th-century concept of spermatozoa*

Despite lacking any formal scientific education, Leeuwenhoek created advanced lenses that enabled him to observe previously unseen things. In 1674, he examined a sample of pond water using his handcrafted microscope. The small, moving creatures—now known as microorganisms-swimming in the water amazed him. It was a revolutionary moment, as it was the initial instance of witnessing living a microscopic life. Many scientists were initially sceptical, but his observations were repeatedly verified, cementing his place in history.[61, 87]

In 1760, Lazzaro Spallanzani, a Catholic priest and scientist, conducted a remarkable experiment with frogs, that challenged the belief in a spontaneous generation from non-living matter. Spallanzani proposed the idea that the combination of eggs and sperm was necessary for creating new life. In order to experiment with his theory, he outfitted a male frog with custom pants to block its sperm during mating. As predicted, the absence of male frog sperm resulted in the eggs remaining unfertilised. The experiment presented convincing proof that reproduction relies on the union of both egg and sperm, thus refuting the notion of spontaneous generation. His work was often controversial but it helped shape the future of biology and germ theory.[148, 159, 161]

Fig 26: *Spallanzani's experiment with frogs*

The first observation of the human ovum by Karl Ernst von Baer in 1827 was a revolutionary discovery that significantly expanded our understanding of human reproduction. In the 1850s, German botanist Nathanael Pringsheim witnessed fertilisation in plants, while

O.Hertwig later observed the same process in humans. These discoveries finally dispelled the myth that tiny pre-formed humans existed inside sperm cells and clarified that reproduction involved the fusion of two distinct cells, transforming our understanding of biology and reproduction.[160]

Gregor Mendel and the Foundations of Genetics

Chromosomes carry genes that control our traits. For example, an organism have two versions (alleles) of a gene for height: notated as 'T' for tallness and 't' for shortness. If the organism has both alleles (Tt), the dominant allele (T) is typically expressed, meaning the organism will be tall. This concept of dominant and recessive traits was first explored by Gregor Mendel through his experiments with pea plants, which laid the foundation for modern genetics. He is renowned as the 'Father of Genetics.'

Gregor Mendel (1822–1884) was an Austrian scientist and monk. His novel research on the inheritance of traits in pea plants laid the foundation for the laws of inheritance.

Fig 27:** Gregor Mendel **(1822–1884)

Mendel discovered that traits are passed from parents to offspring in discrete units, later understood as genes. His work, published in 1866, went unnoticed during his lifetime, but became foundational to the field of genetics in the early 20th century.[219]

When Mendel presented his findings to the scientific community in 1865, many scientists didn't fully grasp their significance. At one point, he sent a copy of his paper to the famous Swiss botanist Carl Nägeli, hoping for recognition. Unfortunately, Nägeli dismissed his work, advising Mendel to conduct experiments with a different plant species that had more complex in heritance patterns, which could have confused Mendel's simple, clear principles. Mendel's work remained obscure until the early 1900s. Long after his death, when three independent scientists rediscovered his research, recognising its revolutionary. Mendelian laws of inheritance are still very much relevant, with some nuances. The laws are 1. The Law of segregation 2. The Law of Independent assortment and 3. The Law of dominance. Regrettably, Charles Darwin was unaware of Mendel's work while he was alive, if he had been, Mendelian genetics would have provided a sound basis for his theory of evolution. [75]

Hugo de Vries introduced the theory of mutation in 1889 in his book 'Intracellular Pangenesis.' He proposed that organisms could pass on changes mutations or Variations caused by mutations could eventually give rise to new species.[249.]

Major Breakthroughs

1) 1953. discovery of double helix structure of DNA - James Watson, Francis Crick, and Maurice Wilkins

2) 1958. Isolating and synthesising DNA in a test tube: Arthur Kornberg and Severo Ochoa

3) 1968. Cracking the genetic code: Marshall Nirenberg, Har Gobind Khorana, and Severo Ochoa

4) 1972., Created the first recombinant DNA molecule by combining DNA from different species. This marked the birth of modern biotechnology: Paul Berg, a biochemist at Stanford University.

5) 1973. Novel feat in genetic engineering by merging a gene from an African frog with the DNA of E. coli, a common bacterium, which successfully produced the frog protein: Researchers from Stanford and the University of California, San Francisco, paved the way for manufacturing human insulin, vaccines, and medications through genetic engineering techniques.

6) 1975. DNA sequencing, which allowed scientists to determine the exact order of nucleotides in DNA.

7) 1988. Genetically engineered crops also made their debut. Eg: Bt -Corn.

8) 1996. Completed the genetic mapping of the mouse genome.

9) 2003. Human Genome Project completed. Marked one of the most monumental achievements in science led by Dr Craig Venter and his team

10) 2006. Discovery of induced Pluripotent Stem Cells (IPSCs): Shinya Yamanaka

11) 2010. Created history by creating the world's first synthetic life form: Craig Venter and his team

12) 2011. Gene-editing tool CRISPR-Cas-9, ('Clustered Regularly Interspaced Short Palindromic Repeats', and 'Cas9' is the associated protein.): Emmanuelle Charpentier and Jennifer Doudna

A major leap in understanding the genetic code came in 1953 when James Watson, Francis Crick, and Maurice Wilkins uncovered the double-helix structure of DNA. This discovery, published in Nature, revolutionised molecular biology. The trio received the Nobel Prize in Physiology or Medicine in 1962 for their work. Sadly, Rosalind Franklin, a British chemist whose x-ray crystallography research was vital in revealing the structure of DNA, did not receive the Nobel Prize because she died in 1958. (Nobel Prize is not awarded posthumously). Her work has received widespread recognition and celebration within the scientific community.[122, 166, 221],

Fig 28: *stalwarts who discovered the double helix structure of DNA*

DNA Synthesis and Decoding genes

in 1958, Arthur Kornberg and Severo Ochoa made another significant breakthrough by isolating and synthesising DNA in a test tube. This achievement won them the Nobel Prize in 1959, and their work led to the discovery of DNA polymerase, an enzyme critical for DNA replication and repair. DNA polymerase has since become a key tool in genetic research and biotechnology.[105]

In 1968, Marshall Nirenberg, Har Gobind Khorana, and Severo Ochoa made another historic contribution by cracking the genetic code. They deciphered how the sequence of 20 amino acids in proteins was encoded in DNA. This accomplishment earned Khorana and Nirenberg the Nobel Prize for Physiology and Medicine.[195]

Recombinant DNA

Recombinant DNA (rDNA) refers to a DNA molecule formed by combining genetic material from two or more different sources. This is achieved by using enzymes to cut and join DNA segments, creating sequences that do not naturally occur together. The resulting recombinant DNA can be inserted into a host organism, such as bacteria or yeast, where it is replicated along with the host's DNA. This technology enables scientists to study genes in detail and has practical applications, such as producing proteins like insulin for medical treatments.

The story of genetic engineering began in 1972, when biochemist Paul Berg at Stanford University made the first recombinant DNA molecule in 1980, Berg received the Nobel Prize in Chemistry for his pioneering work. His discovery opened up new possibilities in science, including the development of gene therapy, where doctors can treat genetic diseases by altering a patient's DNA. Thanks to Berg's early work, scientists now have powerful tools to understand and improve life at the genetic level.[41,121]

Gene Transfer

In 1973, scientists from Stanford University and UCSF made an innovative achievement in genetic engineering. They combined a gene from an African frog with the DNA of E. coli (a common type of bacteria, a work-horse in genetic studies). This bacteria successfully produced a protein of the frog. This experiment opened the door to producing human insulin, vaccines, and other medicines using genetic engineering. This study has become a foundation for modern biotechnology.[239]

DNA sequencing

In 1975, scientists achieved another milestone in genetics by developing DNA sequencing, enabling them to determine the exact order of nucleotides in DNA. This advancement significantly enhanced our understanding of genetic information and its role in living organisms. Genentech, a big name in genetic engineering, was started by Herbert Boyer in 1976. Genentech was the first to make human insulin from bacteria in 1982. This technology's capabilities expanded by 1986, leading to the creation of recombinant vaccines for Hepatitis B and HPV. In medicine, gene therapy and genetic engineering are making major strides in improving patient care.[102, 172, 283,]

Genetically engineered crops

In 1988, genetically changed crops were introduced to farming for the first time. One example is 'Bt corn', a type of corn engineered to produce a natural insect toxin from bacteria called *Bacillus thuringiensis*. This innovation allowed farmers to manage pests without relying on chemical pesticides, marking the start of using genetically modified organisms (GMOs) in agriculture.

In 2003, a unique genetically engineered pet called the GloFish was introduced. These fish contain a gene from coral or jellyfish that makes them glow under ultraviolet (UV) or blue light, giving them bright fluorescent colours. Marketed as GloFish, they became the first genetically changed animals sold to the public, showcasing genetic engineering's potential even in pet breeding.

Genetic engineering allows for the creation of plants that glow. A biotechnology firm in Idaho, Light Bio, has created petunias that glow green in the dark using genetic engineering. The plants contain genes from the bioluminescent mushroom '*Neonothopanus nambi*', which produces light when oxygen reacts with luciferin. These petunias have white flowers that change to a glowing green in the dark.

Fig 29: *Genetically engineered Petunias*

Genome mapping

In 1996, scientists completed the genetic mapping of the mouse genome. This was a significant achievement, as mice share about 85% of their genes with humans, making them valuable for studying human genetic

diseases. Insights from mouse genetics have helped advance medical research and the understanding of disease mechanisms.[45,135, 143, 211, 257]

Human Genome Project

"To grasp the essence of life is to understand that its blueprint is a code—a language that can be read, copied, and edited." *Craig Venter*

The most fascinating stories in the world of genome mapping is the race to decode the human genome in the 1990s, a scientific rivalry that unfolded between two key players: the Human Genome Project (HGP), a publicly funded international effort, and Celera Genomics, a private company led by scientist Craig Venter. Instead of following the HGP's step-by-step process, Venter proposed breaking the genome into random fragments and reassembling them with powerful computer algorithms.[41, 115, 232]

Fig 30: *Craig Venter.*

The competition spurred swift scientific progress. Both teams vied to be the first to publish their results, motivating each other to speed up their efforts. In June 2000, a significant and symbolic truce was declared when President Bill Clinton and UK Prime Minister Tony Blair revealed that both the public and private teams had successfully finished a rough draft of the human genome. While neither team had completely mapped it, they had enough information to make history.[263]

This race not only showcased the spirit of scientific competition but also paved the way for modern genomics and personalised medicine, as the human genome map is now a cornerstone for understanding disease and genetic diversity. In the end, the public project and Celera, both contributed to the final map, and the entire sequence was made publicly available, a victory for global science.[41]

Gene therapy

Gene therapy is a significant advancement in modern medicine that has the potential to treat genetic disorders that were previously incurable. Ashanti De Silva, a four-year-old girl, was the first person to undergo successful gene therapy in 1990. She was born with a condition known as 'Severe Combined Immuno-Deficiency (SCID)', commonly referred to as "*Bubble Boy Disease*." Kids with SCID have no working immune system, making them exceedingly susceptible to infections. Even a simple cold could be deadly. Dr W. French Anderson, a scientist at NIH, was working on an innovative idea called gene therapy. From Ashanti's body, he obtained white blood cells. The team then used an altered virus (retrovirus) to insert the healthy gene into the cells. They inserted the altered white blood cells, now containing the correct gene, back into Ashanti's bloodstream. These altered cells generated the enzyme she lacked, restoring her immune system's functionality. Ashanti's immune system gradually strengthened, enabling her to combat infections. She was no longer required to live in isolation, and her life expectancy, previously uncertain, drastically increased. The way doctors treated Ashanti in 1990 proved a crucial turning point for gene therapy. However, she had to receive genetically engineered white blood cell infusions periodically.

Gene therapy is a promising approach to treat both inherited and acquired diseases. This technology involves introducing a functional gene into a patient's cells to correct or replace a defective gene. For example, in some cancers like leukaemia, gene therapy can alter immune cells to target and destroy cancerous cells more effectively. While there are still challenges, such as ensuring the safety and efficacy of gene therapy continues to offer hope for treating a wide range of diseases.[218]

In 2012, six-year-old Emily Whitehead became the first paediatric patient to receive CAR T-cell therapy for 'Acute Lymphoblastic Leukaemia' (ALL) after standard treatments failed. In this instance, gene therapy changed the patient's immune cells to identify and destroy malignant cells. Following the treatment, Emily achieved complete remission and has remained cancer-free. Her case has been widely recognised as a landmark success in the application of gene therapy for cancer treatment.[268]

Mendelian diseases, named after Gregor Mendel, are caused by mutations in a single gene. Mendelian diseases, such as cystic fibrosis, sickle cell anaemia, and Huntington's disease, follow predictable inheritance patterns based on Mendel's laws. Understanding these patterns is crucial for genetic counselling, diagnosis, and disease management.

Induced pluripotent stem cells (IPSCs).

Stem cells are essential in regenerative medicine because they act as the body's repair system, holding the original DNA blueprint for building various tissues and organs. Stem cells are often present in the bone marrow and can stay dormant until the body requires them to repair or regenerate damaged tissues. Stem cells are the source for all specialised cells in the body, giving them enormous potential for medical treatments.[11]

Japanese scientist Shinya Yamanaka revolutionised stem cell research by discovering induced pluripotent stem cells (IPSCs). These are regular cells, like skin or blood cells, that are reprogrammed to behave like embryonic stem cells. IPSCs have the ability to develop into many types of cells, making them a powerful tool for treating a wide range of diseases. Yamanaka's discovery earned him the Nobel Prize in 2008.[1, 244]

Synthetic life

In 2010, scientist Craig Venter and his team made history in synthetic biology by creating the world's first synthetic life form. They synthesised the complete DNA of Mycoplasma mycoides, a bacterium, and inserted it into another bacterium's empty shell. This new synthetic cell began functioning as Mycoplasma mycoides, essentially creating life in the lab. While this breakthrough opened new doors in science, it also raised ethical and safety concerns. Critics worried about the probable risks of creating synthetic organisms, including the possibility of environmental threats or new diseases. By venturing into synthetic life creation, Venter demonstrated his continued determination to push boundaries in biology, just as he had done with sequencing the human genome. His team's efforts established the groundwork for future advancements in synthetic biology, with diverse applications, including biofuels and pharmaceuticals.

An interesting observation is that they embedded watermarks into the DNA, which included famous quotes and the names of key researchers, effectively labelling the genome as human-made. This process translates sequences of nucleotides (A, T, C, and G) into patterns that convey meaningful information. This is possible because DNA's nucleotide sequence can be designed to represent any kind of information, similar to binary in computers. It is possible to use DNA for data storage. Because of its compactness and stability, DNA can store large amounts of data. Researchers have successfully encoded digital files, like books and videos, into DNA strands. Theoretically, just one gram of DNA can store 215 petabytes of data. The process includes transforming binary code into DNA sequences. Although in its experimental phase and costly, DNA data storage can revolutionise the way we store vast amounts of information.

Gene-editing tools

In 2011, a gene-editing tool called TALENs (Transcription Activator-Like Effector Nucleases) was developed. However, the most powerful and widely used tool today is CRISPR-Cas9. CRISPR stands for '*Clustered Regularly Interspaced Short Palindromic Repeats*', and '*Cas9*' is the associated protein. This system, which comes from the natural immune response of bacteria, can target and cut specific DNA sequences. Cas9 acts like molecular scissors, precisely cutting the DNA at targeted points. CRISPR-Cas9 allows scientists to alter the genes of organisms with remarkable accuracy. Its flexibility has led to breakthroughs in fields like agriculture, medicine, and biotechnology, offering endless possibilities for genetic modification and advancement.[36,123]

The narrative surrounding the invention of CRISPR-Cas9 is one that embodies curiosity, collaboration, and serendipity. During the early 2000s, researchers examining bacterial immune systems observed that bacteria possessed the ability to retain 'viral information' through the storage of DNA fragments in repetitive sequences called CRISPR. The true potential of the CRISPR system only became apparent following the commencement of research by Emmanuelle Charpentier and Jennifer Doudna in 2011.[164]

Emmanuelle Charpentier and Jennifer Doudna, perfected the technology, which earned them the Nobel Prize in 2020. This technology could correct genetic disorders by replacing defective genes with healthy ones. However, this also raises ethical concerns. For instance, changing the genes of crops could affect ecosystems and biodiversity, and there is fear about the unintended consequences of releasing genetically altered organisms into the environment. The possibility of using gene editing for "designer babies" or even reviving extinct species poses deep ethical and philosophical questions about the limits of human intervention . [28, 149, 204.]

Fig 31: *Emmanuelle Charpentier and Jennifer Duodna Receiving Nobel prize for CRISPR gene editing*

Chapter 6

HACKING NATURE

Future: Unveiling the Next Frontier

Looking back at key moments in genetic engineering—from the discovery of DNA in 1869 to the development of CRISPR-Cas9 in 2012—humanity has made remarkable progress in understanding and manipulating life. These advances will have lasting impacts on healthcare, agriculture, and environmental conservation. As we move forward, we must consider the ethical, social, and environmental challenges to ensure responsible use of genetic engineering for a better future.

Artificial Insemination

Artificial insemination, the process of placing semen into the uterus for fertilization, has been practiced for centuries. In the 1770s, Scottish surgeon John Hunter pioneered human artificial insemination, shaping modern reproductive medicine. Hunter was approached by a man unable to consummate his marriage due to a physical condition. Seeing an opportunity, Hunter instructed the husband to collect his semen, which he then carefully introduced into the wife's uterus using a syringe. The procedure succeeded, resulting in a healthy birth—one of the earliest recorded cases of artificial insemination in human history.

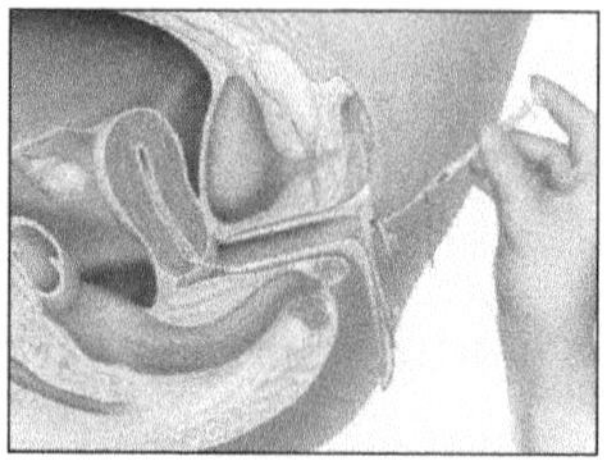

Fig 32: *Artificial insemination*

In Vitro Fertilisation (IVF)

In vitro fertilisation (IVF) is a medical procedure where an egg is fertilised by sperm outside the body, typically in a laboratory dish. The resulting embryo is then implanted into the uterus to establish a pregnancy. The first successful IVF birth was Louise Brown in 1978, sparking debates but eventually gaining acceptance. Today, IVF is widely used to help individuals and couples facing infertility, with advanced techniques like sperm microinjection further improving its success rates.

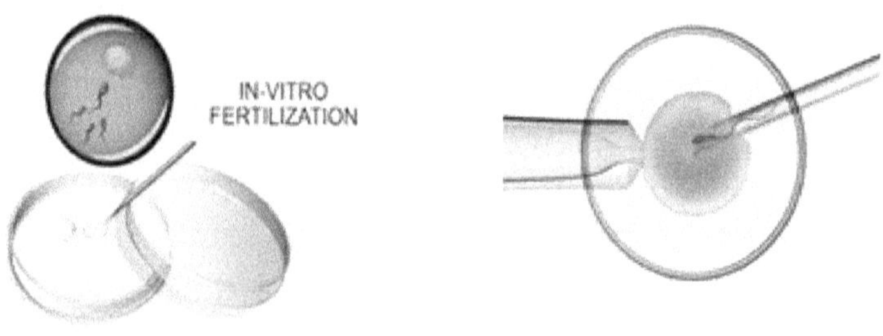

Fig 33: *A- In-Vitro Fertlisation* ***B-*** *Microinjection of sperm into the egg,*

Fig 34: *A: British physiologist Robert Edwards, left received Nobel prize for his pioneering work B: The world's first test tube baby Louise Joy Brown is celebrating her 40th year with her children*

IVF was seen by many within the Church as a form of " *Playing God*" because it separated conception from the natural act of sexual union. The Church officially condemned IVF in 1987 through a document called *Donum Vitae* (The Gift of Life), issued by the Congregation for the Doctrine of the Faith under, then Cardinal Joseph Ratzinger, later Pope Benedict XVI. The story of IVF's early years is a vivid example of the clash between scientific progress and religious ethics, one that has spurred deep reflection on what it means to create life.

Kanupriya Agarwal, India's inaugural IVF baby, also known as Durga, entered the world on October 3, 1978, a few months following the birth of the world's first

IVF baby. Dr Subhash Mukhopadhyay, a gynaecologist and reproductive biologist based in Kolkata, conducted the procedure. Notwithstanding, his pioneering accomplishment, Dr Mukhopadhyay was met with substantial scepticism and a dearth of acknowledgement from the Indian medical establishment during that specific period. Recognition for his contributions came long after his passing. Despite the initial doubts, IVF has gained wide-spread acceptance and now has centres even in rural areas.

Cloning

Cloning involves the creation of an exact replica of an organism or cell. Two primary categories of cloning exist: gene cloning and whole-organism cloning. The process of gene cloning entails generating multiple replicas of a particular gene. This is often done by inserting the gene of interest into a vector, such as a bacterial plasmid or a viral genome. The vector is then introduced into a host cell, where it replicates along with the inserted gene. This process can be used to produce large quantities of proteins or enzymes.

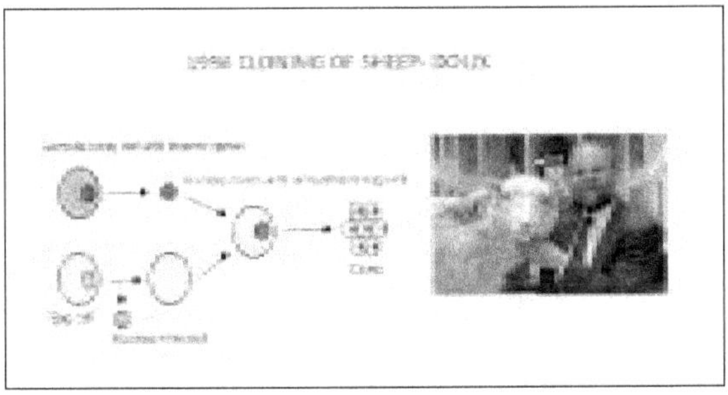

Fig 35: *Whole organism cloning; Sheep Dolly and creator Dr Ian Wilmut.*

The process of whole-organism cloning results in the creation of a genetically identical organism. This is typically done by transferring the nucleus from a somatic cell (a non-reproductive cell) of one organism into an egg cell that has had its nucleus removed. The resulting embryo is then implanted into a surrogate mother, where it can develop into a clone of the original organism. This process is known as Somatic Cell Nuclear Transfer (SCNT) and was used to create 'Dolly the cloned sheep', the first cloned mammal.

Medicine, agriculture, and research all stand to benefit from the potential applications of cloning. Yet, it also raises ethical dilemmas, particularly in relation to the cloning of humans.

The combination of genetic engineering organ transplantation, and longevity research offers new ways to address challenges like organ failure and age-related diseases. Advances in gene expression optimisation, genetic screening, and Organ engineering could revolutionise healthcare, improving treatment outcomes for patients worldwide.

Xenotransplantation involves transplanting organs from genetically engineered pigs to humans, solving the organ donor shortage. While this approach demonstrated in the case of David Bennett Sr.'s heart transplant faced certain challenges.

Presently, researchers modify pigs to create organs that are less likely to trigger immune rejection in humans. Despite certain challenges, it offers hope for patients waiting for life-saving organ transplants. Genetic engineering, though controversial, continues to advance, offering solutions to medical problems.[272]

Genetic Research on Longevity

Advances in genetic research, such as identifying ageing-related genes like DAF-2, offer new possibilities for extending human lifespan and improving health. By targeting genes involved in ageing and cell signalling, scientists aim to develop treatments that could delay age-related diseases and increase longevity. For example, a study by Dr Alexei Maklakov's team doubled the lifespan of round worms by silencing the DAF-2 gene, highlighting the possibility of genetic manipulation in longevity research. Although more work is needed to apply this to humans, this area of study is promising.

Telomeres and Cellular Ageing

Telomeres, the protective caps at the ends of chromosomes, play a vital role in maintaining chromosome stability. As cells divide, telomeres shorten, which are linked to ageing and the onset of age-related diseases. Telomerase, an enzyme that extends telomeres, helps certain cells, like stem and germ cells, continue dividing. However, as telomerase levels decline with age, telomeres shorten, contributing to cellular ageing. This relationship between telomeres and ageing is a key focus in longevity research, particularly for its role in preventing age-related diseases.[21, 98, 125, 265]

CRISPR-Cas9 technology and gene editing

CRISPR-Cas9 technology has revolutionised genome editing and has made it much more accessible and efficient compared to previous methods. CRISPR-Cas9, acts like molecular scissors, allowing scientists to cut and paste specific genes/ DNA strands within an organism's genome, including in embryos. CRISPR-Cas9 holds great promise for therapeutic applications, including the correction of genetic diseases. Nevertheless, it raises significant ethical concerns, particularly when it comes to editing the human germline. Germline genome editing involves making changes to the DNA of sperm, eggs or embryos, which would result in heritable genetic modifications that would be passed on to future generations[5, 64.]

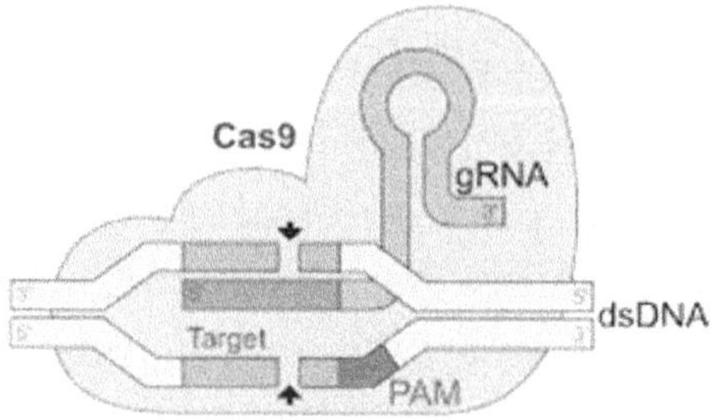

Fig 36: *Gene editing using CRISPER-Cas9*

Pre-implantation Genetic Screening and Diagnosis (PGS&PGD)

PGS and PGD are techniques used with in vitro fertilisation (IVF) to screen embryos for genetic issues before implantation. PGS checks for chromosomal abnormalities to improve implantation success and reduce miscarriage risks. PGD tests for specific genetic disorders, like cystic fibrosis or Huntington's disease, help parents select embryos free from genetic mutations/ disease-causing genes for implantation. These techniques provide valuable tools for preventing hereditary diseases and improving reproductive outcomes.[64]

The process of PGS and PGD involves the following steps:

1. Multiple eggs are retrieved from the ovaries and fertilised with sperm in the laboratory through IVF.
2. The resulting embryos are allowed to grow for a few days until they reach the 6- to 10-cell stage.

3. At this stage, one or more cells are removed from each embryo for genetic testing, either through a biopsy or by extracting a small portion of the embryo.
4. The genetic material from the biopsied cells are analysed using various techniques, such as polymerase chain reaction (PCR), fluorescent in situ hybridisation (FISH) or next-generation sequencing (NGS), depending on the type of screening or diagnosis required.
5. Based on the results of genetic testing, embryos free of the targeted genetic abnormalities are selected for implantation into the woman's uterus.
6. Any remaining embryos that do not carry genetic abnormalities maybe discarded or cryopreserved for future use.

PGS and PGD offer couples the opportunity to increase their chances of having a healthy baby while minimising the risk of passing on genetic disorders to their offspring. However, it's important to note that these procedures are not without limitations and ethical considerations.[5, 118, 130,]

The consequences of germ-line genome editing in humans are complex and far-reaching. While it could theoretically prevent or eliminate genetic diseases, there are concerns about unintended off-target effects, the long-term safety and efficacy of the procedure, and the likelihood of misuse, including the creation of so-called "designer babies" with enhanced traits.

Due to these ethical and safety concerns, many countries have regulations or outright bans in place prohibiting germline genome editing in humans. These regulations are intended to ensure that any research involving genome editing is conducted responsibly and ethically, with careful consideration given to the possible risks and benefits.

While research on germline genome editing goes on in non-human organisms, the prospect of using this technology in humans remains a subject of ongoing debate and discussion within the scientific community and society at large. Ethical, legal, and societal considerations will play a crucial role in determining the future trajectory of genome editing technologies and their applications in humans.[50]

The announcement by He Jiankui in Nov 2018 regarding the birth of genetically modified babies sparked widespread condemnation and raised serious ethical concerns within the scientific community and society as a whole. He Jiankui claimed to have used CRISPR-Cas9 technology to edit the genomes of embryos to confer resistance to HIV, and the resulting births of Lulu and Nana (twin girls) represented the first documented case of germline genome editing in humans. The experiment was widely criticised for several reasons. First and foremost, it was conducted without proper ethical oversight or transparency, raising concerns about the informed consent of the parents involved and the long-term health effects on the children. Additionally, the scientific community expressed scepticism about the safety and efficacy of the gene editing technology used, as well as the unintended consequences and risks associated with making heritable genetic modifications. The incident underscored the urgent need for clear guidelines and regulations governing the ethical conduct of research involving human germline editing. Many countries have since strengthened their regulations or issued outright bans on germline genome editing in humans, emphasising the importance of responsible and ethical research practices in this fast advancing field.[57, 101]

While the desire to prevent genetic diseases and improve human health is understandable, the risks and ethical considerations associated with germline genome editing necessitate careful deliberation and robust oversight. The case of He Jiankui serves as a cautionary tale about the dangers of crossing ethical boundaries in the pursuit of scientific progress, and highlights the importance of upholding ethical principles and protecting the well-being of future generations.

Artificial wombs (Bio-bags)

The development of artificial wombs, also known as exo-wombs or bio bags, represents a significant advancement in neonatal care and reproductive technology. By providing an environment that mimics the conditions of the womb, artificial wombs offer substantial benefits for premature infants and could revolutionise the way we approach neonatal care.

One of the primary advantages of artificial wombs is their ability to support the growth and development of premature babies outside the mother's body. Premature birth can pose significant health risks for infants, including respiratory problems, neurological complications, and developmental delays. By allowing premature infants to continue their development in an environment that closely resembles the womb, artificial wombs have the ability to reduce the incidence of these complications and improve outcomes for preterm babies. Furthermore, artificial wombs could enable medical interventions and treatments that are not feasible 'in utero' or immediately after birth. For example, surgeons could perform procedures to correct birth defects or treat medical conditions while the foetus is still in the artificial womb, reducing the risks associated with traditional postnatal surgery.

Additionally, artificial wombs may offer new possibilities for assisted reproduction and fertility treatment. For couples struggling with infertility or individuals unable to carry a pregnancy to term, artificial wombs could provide an alternative option for gestating embryos and bringing pregnancies to full term.[199, 273.]

However, the development and implementation of artificial wombs also raise important ethical and social considerations. Questions about the appropriate use of this technology, the impact on traditional notions of pregnancy and childbirth, and concerns about the commodification of reproduction will need to be addressed as artificial wombs become more sophisticated and widespread.[59, 103]

Overall, artificial wombs hold great promise for improving neonatal outcomes and expanding reproductive options, but careful consideration of the ethical, social, and legal implications will be essential as this technology continues to advance.[103]

Creating "Designer Babies"

"A society that allows cloning has, whether it knows it or not, tacitly said yes to converting procreation into manufacture and to treating our children as pure projects of our will."

Leon Kass, bioethicist

Creating a "Designer Baby" refers to using genetic technologies to select or change an embryo's traits before birth. Advancements in gene editing and reproductive technologies have made it theoretically possible. Here's an overview of the steps that might be involved in creating a designer baby using current or emerging technologies:

The first step is to create embryos using IVF. Multiple embryos are created. Perform genome analysis and subject it to precise editing using CRISPR-Cas9, a revolutionary gene-editing tool, to precisely edit genes. To create a designer baby, specific genes associated with desired traits—like height, intelligence or physical appearance—would need to be identified. Using CRISPR-Cas9, scientists could cut out or replace sections of the embryo's DNA to change genes linked to those traits. This could involve correcting a genetic defect or even enhancing certain qualities. Successfully edited embryos can be implanted in the mother's uterus.

However, many genetic traits, like talent, height, intelligence, etc., are vastly complex and not fully understood.

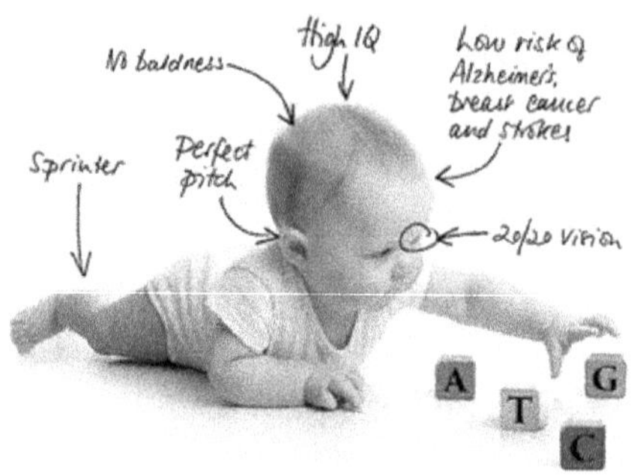

Fig 37: Designer Baby

The scenario outlined presents a fascinating glimpse into the future of reproductive technology and genetic engineering. While the idea of creating "Designer Babies" with specific desired traits may seem appealing, it also raises a host of ethical, social, and practical concerns that must be carefully considered.

On the one hand, advances in reproductive technology could offer new opportunities for individuals and couples facing infertility or genetic disorders to have healthy children. Technologies like in vitro fertilisation (IVF), pre-implantation genetic screening, and gene editing might reduce the risk of genetic diseases and improve the overall health and well-being of future generations. Nevertheless, the prospect of "designer babies" also raises serious ethical questions about the implications of manipulating human genetics. Concerns about eugenics, inequality, and the commodification of human life come to the forefront when considering the consequences of selecting specific traits in embryos. There is also the risk of unintended consequences or unforeseen side effects from genetic modifications, as well as the danger of discrimination and social division based on genetic status.

Once the bio bag becomes sufficiently advanced to accept the embryo and develop a child, the need for a mother's uterus dims. This means the authorities can produce humans to the whims of the ruling class, from page boys to scientists and rulers. This could spawn a dystopian society of disparity and inequality.

Furthermore, the idea of a totalitarian government controlling the reproduction of its citizens is deeply troubling and echoes dystopian visions like those depicted in Aldous Huxley's, "*Brave New World*". Such scenarios underscore the importance of legal regulations, and public dialogue to ensure that advances in genetic engineering are used responsibly and follow fundamental principles of human dignity, autonomy, and equality.

Ultimately, the future of reproductive technology and genetic engineering will depend on how society navigates these complex ethical and moral challenges. It will require careful consideration of the benefits and risks, as well as ongoing dialogue and engagement with diverse stakeholders to ensure that scientific advances are pursued in ways that promote respect for human rights.

In the end, our choices today will reverberate far into the future, shaping the destiny of life on Earth. Let us approach this responsibility with reverence, courage, and a steadfast commitment to the common good. For in the journey of decoding life, our collective wisdom and moral integrity will be our most precious guides

Nanotechnology

"Nanotechnology is manufacturing with atoms."

William Powell,

Nanotechnology could transform human life in ways we're just beginning to understand. As technology grows faster than ever, nanotechnology stands out for its potential to solve some of our biggest global challenges. Nanotechnology provides new solutions, from medical breakthroughs to addressing pollution. Even though it's still developing, the rapid progress hints at a future where humanity can use the unique powers of atoms and molecules to create innovations that were once science fiction.

Nanotechnology involves manipulating materials on a minuscule scale of one-billionth of a meter. At this size, materials exhibit unusual behaviours, such as increased strength, decreased weight, heightened reactivity or improved electrical conductivity. Nanotechnology enables scientists to construct miniature machines, modify atomic-level structures, and develop specialised materials. Because of its versatile applications, nanotechnology has the potential to create an impact across various sectors, such as healthcare, energy, and environmental sustainability.

Nanomedicine

'Nanotechnology in medicine is going to have a major impact on the survival of the human race."

Bernard Marcus,

Nanotechnology is already changing healthcare, and its future potential could be revolutionary. Nanomedicine uses nanotechnology to diagnose, treat, and monitor health conditions in ways that were previously impossible. With this precision, treatments can target specific cells or molecules, which makes them more effective and reduces side effects.

One promising example is targeted drug delivery. Nanoparticles can deliver medicine directly to cancer cells, minimising harm to healthy cells. In the future, "smart" nanoparticles could release medicine only when needed, responding to certain conditions like changes in acidity or specific proteins. This could lead to personalised treatments for diseases like cancer or Alzheimer's, improving quality of life and extending lifespans.

Nano-sensors also hold promise for detecting diseases early. Tiny sensors in the bloodstream could identify subtle biochemical changes, catching diseases like cancer or heart disease early. These sensors could even work with wearable devices to monitor health in real-time, making healthcare more about prevention than treatment.

Many applications, like nanobots for surgery or more advanced forms of tissue regeneration, are still under research and development. The field is advancing, and with continued progress. Nanotechnology is expected to play critical role in the future of healthcare.

One example of nanoparticles used in healthcare is lipid nanoparticles (LNPs), which have been crucial in the delivery of mRNA vaccines, such as the COVID-19 vaccines from Pfizer-BioNTech and Modern. These tiny particles, made of lipids (fat-like molecules), are used to transport mRNA molecules into cells. Lipid nanoparticles have become a safe, effective way to deliver genetic materials for and adopt policies that promote environmental stewardship and social equity. By working together, we can create a more sustainable future for ourselves and generations to come. Vaccine development hold promise for future therapies, including cancer treatments and gene therapies, by enabling precise delivery to target cells

Enhancing Human Abilities

Nanotechnology might not just treat diseases; it could also enhance our physical and mental abilities. Nanobots, tiny machines that could move through the body, are being developed to repair tissues or clear blocked blood vessels. Some researchers even think that nanobots could help improve brain function by connecting directly to neurons, which could enhance memory, processing speed, and problem-solving skills. This could mark the advent of "augmented humanity," where technology and biology merge seamlessly to expand the limits of human potential.[18]

Nanotechnology could also make prosthetics and implants more comfortable and long-lasting. With nanomaterials, artificial limbs and implants would be lightweight, durable, and more compatible with the human body. Advances in neural interfaces could even allow people to control prosthetic limbs with their thoughts, transforming the lives of those with disabilities. However, many applications, like nanobots for surgery or more advanced forms of tissue regeneration, are still under research and development. The field is advancing rapidly, and with continued progress. Nanotechnology is expected to play critical role in the future of healthcare.

Environmental Impact and Sustainability

Nanotechnology isn't just about health and enhancement; it could help tackle environmental challenges, too. In the energy field, for instance, nanotechnology could change renewable energy sources like solar power. Nanomaterials like graphene and carbon nanotubes could make solar cells cheaper, more efficient, and flexible. They also have the potential to improve batteries, making them last longer, charge faster, and degrade less. This could lead to a more sustainable energy infrastructure.[117]

Nanotechnology can significantly change manufacturing, especially with new, advanced materials. Traditional materials have limits, but nanomaterials can be designed to be stronger, more flexible, and more durable. For instance, carbon nanotubes are super strong and lightweight, making them ideal for building lighter vehicles, from airplanes to spacecraft.

3D printing with nanomaterials allows us to create complex structures that we couldn't build before. Imagine buildings that can repair themselves or clothes that change colour or adjust to temperature automatically. Nanotechnology could make products more adaptive, efficient, and environmental friendly, changing how we interact with the material world.

The Future of Nanotechnology and Humanity

As we look to the future, nanotechnology's impact on humanity is both promising and complex. It could completely change healthcare, giving us the power to fight diseases and enhance human capabilities. It could help protect our environment and lead to cleaner energy solutions. And it could change manufacturing, allowing us to create stronger, more adaptive materials.

Nanotechnology is on the edge of transforming how

we live. As scientists and engineers discover new ways to use it, we're approaching a future where life could look very different. This journey will need both creativity and caution as we navigate the possibilities and risks of this powerful technology. As nanotechnology advances, it's reshaping our vision of humanity's future—a future where health, sustainability, and technology work together to redefine what's possible.[16, 23, 62, 21]

Anthropogenic Self-threat to humanity

"We declare, with more than 11,000 scientist signatories from around the world, clearly and unequivocally that planet Earth is facing a climate emergency."

World Scientists' Warning to Humanity (2019).

The human species, a recent entrant in Earth's history, has made a significant imprint on the planet. From the dawn of agriculture to the industrial revolution, human activities have transformed ecosystems, altered climate patterns, and depleted natural resources. Population growth, resource consumption, and climate change are the key drivers of environmental degradation. The remarkable growth of the human population plays a significant role in putting more pressure on our planet's resources. In just the last hundred years, the global population has jumped from around 1.6 billion to over 8.19 billion!. This swift increase has been driven by exciting advancements in agriculture, medicine, and public health, which have together raised life expectancy and lowered mortality rates. The distribution of population is another serious factor, with densely populated regions facing acute challenges such as water scarcity, air pollution, and waste management.

The global economy is largely based on a linear model of production and consumption, where resources are extracted, used, and then discarded as waste. This unsustainable pattern has depleted finite resources and generated vast amounts of pollution. The growing affluence of many societies has also contributed to increased consumption, further exacerbating resource scarcity. The burning of fossil fuels for energy production has released vast quantities of greenhouse gases into the atmosphere, leading to global warming and climate change. The consequences of climate change are far-reaching, including rising sea levels, extreme weather events, ocean acidification, and loss of biodiversity.

The human footprint on the planet is undeniable. The consequences of our actions are becoming increasingly apparent. Population growth, resource consumption, and climate change are interconnected challenges that require urgent attention. Addressing these issues will require a fundamental shift towards sustainable practices, including reducing our reliance on fossil fuels, conserving resources, and protecting ecosystems. Considering the above mentioned snags and pattern of our actions, the important thought that comes into the mind is 'Malthusian theory of population.' If we replace 'food' with 'air and water,' the theory becomes remarkably relevant to modern times. Are we on the verge of an Anthropogenic extinction? Earth, being a self-sustaining planet, has survived five great extinction events and changed the direction and progression of life. We being the product of the fifth extinction, do we still believe that the universe is made for us? Do we think we can exploit and deplete it without realising that such actions will ultimately lead to our peril?

It is imperative that we develop innovative solutions and adopt policies that promote environmental stewardship and social equity. By working together, we can create a more sustainable future for ourselves and generations to come.

Chapter 7

DIGITAL WORLD, AI AND FUTURE OF HUMANITY

"By far, the greatest danger of Artificial Intelligence is that people conclude too early that they understand it."

Eliezer Yudkowsky

"Artificial intelligence will be the ultimate tool, but it will also be the ultimate weapon."

Demis Hassabis

The concept of a machine that could act like a human was imagined by British mathematician Alan Turing. He came up with a concept called the "Turing Machine." Instead of asking, "Can machines think?", Turing suggested a test: If a person interacts with a machine and can't tell whether they're talking to a human or a computer, then the machine is having passed the "Turing Test." This test is still used today to measure how advanced artificial intelligence is. The Turing Test has faced criticism for its focus on linguistic ability, rather than measuring true intelligence or consciousness. However, in contemporary AI, machine learning, deep learning, and artificial general intelligence are given priority.[188, 222, 260]

The Atanasoff-Berry Computer (ABC), created between 1937, and 1942, was one of the earliest digital computers. It solved complex math equations, which are commonly needed in fields like physics and engineering. The ABC laid the groundwork for modern computers.

A major step forward in computing was the creation of ENIAC, one of the first general-purpose computers. Built in 1945 during World War II by John Mauchly and J. Presper Eckert at the University of Pennsylvania, ENIAC was enormous, weighing around 30 tons and filling a space of 1,800 square feet. It could be programmed to handle many kinds of calculations, making it a huge achievement in computing.[228]

In 1947, Bell Labs invented the transistor, which was a big breakthrough for computing. Transistors replaced the bulky vacuum tubes used before, making computers smaller, faster, and more energy-efficient. Around this time, programming languages like FORTRAN and COBOL were created, making it easier for people to interact with computers. In the 1970s, the invention of the microprocessor changed everything. Microprocessors allowed for the creation of personal computers (PCs), making computers available to everyone. This new era of personal computing started with machines like the Altair 8800, Apple II, and IBM PC.

The term "Artificial Intelligence" (AI) was first used

in 1956 by John McCarthy, an American computer scientist, at a conference at Dartmouth. This event was a turning point in computing, kick starting research in AI and eventually leading to advanced tools like machine learning (where computers learn from data) and language processing (where they understand human language).[188,222]

In 1965, Moore noted a doubling of transistors on computer chips every few years without rising costs. Moore's Law predicted rapid growth in computing power. This idea set a tech world standard for rapid technological advancement. Despite physical challenges, Moore's Law impacted electronics for years.[114.] Moore never named or intended his observation as a law. It was his friend, Dr. Carver Mead from Cal-Tech, who coined the term "Moore's Law". Later on PC and AI research spurred rapid technological advancement. [100, 177,188,] Small organisations and individuals can now participate in research. Easy code and data sharing online fuelled AI development. Open AI thrived on this sharing. Graphics processing units (GPUs) were also developed, which enabled computers to handle complex AI tasks like deep learning. The internet, which started as a military-funded research network, eventually became the global network we know today-the World Wide Web (www)

Deep learning, a type of AI, mimics the human brain to recognise patterns in data. It uses multiple network layers to learn features and process vast amounts of data. This enables tasks like image recognition, speech processing, and translation, driving AI advancements in autonomous vehicles, healthcare, and education. Deep learning has revolutionised AI, powering breakthroughs in computer vision and language processing. IBM's Deep Blue beating Garry Kasparov in 1977, highlights AI's strength in complex tasks. [90, 150]

Machine learning, a branch of AI, helps computers learn from experience, like humans. Using neural networks, it processes data to improve tasks like language understanding, search engines, and facial recognition. It continuously improves by analysing massive data from sources like social media and business records. Machine learning powers voice assistants, fraud detection, disease diagnosis, and self-driving cars, making computers smarter and problem-solving easier.[10, 12, 38, 3]

AI is transforming society through machine learning, robotics, and language processing, analysing vast data in seconds. In 1985, David Deutsch proposed that quantum computers could simulate any physical system, solving problems beyond classical computers. Using qubits instead of binary logic, they process information much faster, enhancing AI in medicine, energy, and industry. Google, in partnership with NASA and Oak Ridge, demonstrated quantum supremacy, solving in seconds what would take supercomputers thousands of years

AI can analyse patient data to identify patterns and recommend personalised treatments. In the future, it may assist in drug discovery. IBM Watson already provides AI-driven cancer treatment recommendations. AI in healthcare enhances early disease detection and improves lifespan.[2, 71, 187, 280]

Autonomous systems of AI power self-driving cars. This can be extended to drones, robots and self-navigating ships. AI, especially Generative AI is making inroads into creative art forms like music, art, paintings and literature. These systems can generate material that is almost akin to human creativity. AI-driven education platforms could revolutionise both traditional and remote learning environments.[171, 189]

AI's rise raises ethical concerns about bias, privacy, etc. Job displacement from automation is a concern, despite productivity gains. Optimists believe AI will change, not eliminate, jobs, especially those needing creativity and complex decision-making.[3, 193,]

Narrow AI: AI programmed for specific tasks (e.g., image recognition, chatbots, recommendation systems). It excels in predefined functions but lacks general reasoning.

General AI: Hypothetical AI that can learn, reason, and perform any intellectual task like a human. It would require advanced programming for adaptability and self-learning.

Super AI: A theoretical AI surpassing human intelligence, capable of independent thinking, creativity, and problem-solving beyond human ability. It would involve breakthroughs in AI algorithms, hardware, and cognitive modelling.

AI has not yet reached General Intelligence and remains at the Narrow Intelligence stage. To match human intelligence, it must advance to General Intelligence, and to surpass it, AI must reach Super Intelligence. This depends on research progress, cognitive breakthroughs, and technological advancements.

While some scientists doubt AI can ever replicate the human brain, General Intelligence remains theoretical. Artificial Superintelligence and Technological Singularity are still hypothetical concepts. [44, 74, 91, 191, 280]

Virtual and Augmented Reality

Wearing a VR headset transports you into a 3D virtual world, from games to real-world simulations. Using hand controllers, you can move and interact. VR is used in gaming, entertainment, virtual tours, and professional training for pilots, doctors, and the military.[258, 266, 267]

Apps like Google Earth and 360-degree tours let you explore museums and landmarks like the Louvre and the Eiffel Tower as if you were there.[124]

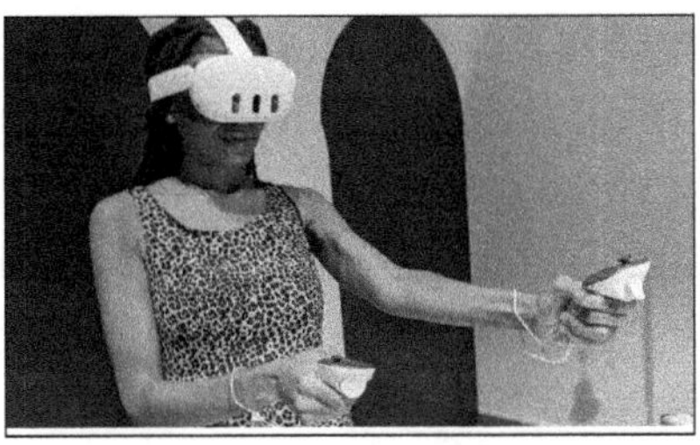

Fig 38: *VR Head Set*

Augmented Reality (AR) overlays digital information onto the real world, like showing directions on roads or placing virtual furniture in your room. It's used in games, navigation, education, and even surgery. Both AR and VR are advancing briskly and may become part of daily life. Mixed Reality (MR) combines elements of both for an even more immersive experience.[13, 20, 24, 151, 175]

The term "metaverse" was first coined by author Neal Stephenson in his 1992 science fiction novel *Snow Crash*, where it referred to a fully immersive, shared virtual reality space. In Stephenson's vision, users, represented by avatars, could interact with each other and digital content in a simulated environment. Technologists have since taken up this concept, working to realise Stephenson's vision.[182, 183]

The metaverse is like the next step in the evolution of the internet. It brings together virtual, augmented, and real-world experiences into one connected digital world. In the metaverse, people can interact with each other and with virtual objects in a 3D environment that feels more realistic than ever.

The metaverse is a continuous virtual world where you can keep your digital identity and assets across platforms. It’s used for gaming, education, work, and healthcare, enabling virtual doctor visits and remote surgeries. Access requires advanced tech like VR headsets and fast internet, which may limit availability.[165]

Digital revolution has completely altered our attitude

and behaviour. It adversely affected our cognitive abilities. Now we are almost bound to our smart phones, which can do multitude of tasks. It has created a 'digital dementia.' Memory recall, navigation, critical analysis, problem solving to art and writing are certain tasks which the AI can take care of. These are going to make people passive consumers rather than active thinkers, which may cause cognitive atrophy. However, there are experts who are optimistic that AI can enhance our cognitive capabilities.[72, 173, 264]

Superintelligent AI

"The development of full artificial intelligence could spell the end of the human race."

— Stephen Hawking

Stephen Jay Gould used his idea 'replaying the tape of life', to argue against the notion that evolution follows a predetermined path toward complexity or intelligence. Instead, he believed that evolution is shaped by contingency-unpredictable events that steer life in unexpected directions. In this view, human existence is not an inevitable result of evolution but rather a fortunate accident of history. Small, random events-such as mutations, extinctions, and environmental shifts-could lead to entirely new species, ecosystems, and possibly no human-like intelligence at all.[93]

Ray Kurzweil, author of 'The Singularity Is Near',

believes that artificial intelligence follows a pattern similar to Moore's Law (the observation that the number of transistors on a chip doubles every 18–24 months). Kurzweil is a public advocate for the ‘futurist’ and ‘trans-humanist’ movements. He predicts that technological and biological advancements will speed up so rapidly and profoundly that human life will undergo an irreversible transformation. This moment, which he refers to as the "Singularity," is expected to occur by 2045.[142]

Kurzweil also predicts that nanobots will play important roles in health science, curing diseases, and extending lifespans. Virtual Reality (VR), Augmented Reality (AR), and the metaverse could become more immersive and akin to reality, not only in entertainment but also in intimate relationships.

Kurzweil proposes that as we replicate and improve upon the structures of the human brain within AI, consciousness will emerge naturally. He theorises that AI will ultimately blend with human intelligence, resulting in a hybrid model where humans integrate with AI. through brain-machine interfaces. This integration would, according to Kurzweil, bring about a form of consciousness in machines that could be indistinguishable from human consciousness. He views this as part of humanity’s evolution, envisioning a future where humans and AI coexist and merge. Kurzweil is optimistic about the development of AI and its benefits to humanity.

His predictions have an accuracy rate of 86%. For example, in 1990, he predicted that a computer would defeat a chess champion by 2000. This prediction came true in 1997 when the computer Deep Blue defeated Garry Kasparov, then reigning champion.

Fig 39: *Ray Kurzweil and Nick Bostrom*

Some researchers in AI ethics, such as Stuart Russell and Nick Bostrom, believe that while AI could follow ethical rules, this does not equate to having a conscience. Machines can be programmed to follow moral guidelines, but these would be based on human-coded algorithms, lacking the subjective experience of right and wrong. The development of moral reasoning in AI would likely be driven by utility functions-algorithms optimised to achieve specific outcomes. Bostrom, author of *Super-intelligence: Paths, Dangers, Strategies*, acknowledges that we may never know if AI truly becomes conscious. This is due to the "hard problem" of consciousness, which questions whether we can determine or even understand another entity's subjective experience. For Bostrom, the moral and existential implications of 'Superintelligent AI' hinge on alignment, safety, and control, rather than consciousness itself.[25, 222]

Another significant concern is whether 'Super intelligent AI' will uphold the ethical standards and values that we cherish. Without careful alignment, AI might disregard human welfare entirely. Experts like Nick Bostrom have warned that if AI's objectives diverge from ours, it could pose an existential threat to humanity, leading to catastrophic consequences. Bostrom contends that the creation of super intelligence could be humanity's most consequential and risky invention, offering immense benefits but also posing existential dangers.

The development of human enhancement technologies, such as CRISPR-Cas9 based gene editing and brain-computer interfaces, could redefine what it means to be human. These enhancements raise concerns about fairness, equality, and the possibility of creating a divided society, where enhanced, individuals hold an advantage over non-enhanced. Michio Kaku, a prominent theoretical physicist, also believes that technological advancements-particularly in AI and biotechnology-will greatly benefit humanity, offering solutions to global challenges such as climate change, disease, and space exploration. He advocates for space exploration and the colonisation of other planets as crucial steps for the survival of humanity. While optimistic, Kaku emphasises the importance of responsibly managing these technologies. (Michio Kaku: books: '*The future of the mind*'(2014); '*The Future of Humanity*' (2018)

Chapter 8

CAN ARTIFICIAL SUPER INTELLIGENT MACHINES DEVELOP CONSCIENCE AND CONSCIOUSNESS?

"Consciousness is the one thing in this universe that cannot be an illusion."

— *Sam Harris*

"How it is that anything so remarkable as a state of consciousness comes about as a result of irritating nervous tissue, is just as unaccountable as the appearance of the Djinn when Aladdin rubbed his lamp in the story".

Thomas Henry Huxley

This is the most important question scientists, neurologists, psychologists, thought leaders, philosophers, and theologists ask. Each of them holds a unique viewpoint on AI, particularly its potential for super intelligence. It arouses the interest of both, 'knowledgeable' and 'not knowledgeable', in the subject.

Before we peep into the topic, it's crucial to grasp the distinctions between conscience, consciousness and sentience. Mind is the centre of many faculties conscience, consciousness, emotions, thinking, perception, memory, and more. Consciousness, often defined as awareness of the external world and self, is universally accepted but remains a deeply debated topic. The origin, seat, and function of consciousness have puzzled philosophers and scientists for millennia, from pre-Aristotelian times to the present. The study of consciousness has long been a central question in philosophy, psychology, and neuroscience. Ancient philosophers such as Plato and Aristotle pondered the nature of the soul and mind, seeking to understand what gives humans the ability to think, feel, and perceive. The 17th-century philosopher René Descartes famously declared, "Cogito, ergo sum" (I think, therefore I am), emphasising consciousness as the core of human identity and experience. He introduced a dualistic view that separated the mind and body as distinct entities, with the mind controlling the body's rational soul. This 'Cartesian dualism' was challenged, and modern views often lean towards monism, which posits a unified reality. Physicalists argue that the mind arises from matter. Idealists claim that thought is primary. Neutral monism suggests both mind and matter stem from a deeper essence.[60, 109]

According to modern Darwinian synthesis, life diversifies because of random mutations that give some organisms advantages in survival and reproduction. Over time, these advantages accumulate, leading to the emergence of new species. Evolution has no specific "goal" or "direction," but the process naturally generates diversity through the pressures of environmental adaptation. Self-awareness and consciousness could have risen because of evolutionary pressures. It implies that consciousness might offer advantages for survival and that even lower-ranked animals have a lesser degree of consciousness. In The Descent of Man, Charles Darwin wrote: "The difference in mind between man and the higher animals, great as it is, is certainly one of degree and not of kind."[35, 81]

Thomas Nagel argues that evolution may have specific goals, such as the development of consciousness. These goals may not be apparent from a purely biological standpoint. Nietzsche proposed that human conscience evolved to regulate behaviour in more complex societies rather than solely for survival. He famously stated, "Man is the animal that must overcome itself."[180.]

While the human mind is considered the most complex object we know, it's true nature remains elusive. Questions about the relationship between the mind and brain or whether animals possess minds, are still unresolved. A key question is whether the mind evolved through natural selection.

Some scientists suggest that consciousness evolved as

a tool to help humans survive better. Antonio Damasio, a well-known neuroscientist, argues that being self-aware helps us make complex decisions and interact with our surroundings more effectively. By being conscious, we can think ahead to avoid future dangers, learn from past experiences, and work together in social groups. These abilities give humans a survival advantage because we can plan, adapt, and cooperate more efficiently.

Neuroscientist, Robert Sapolsky points out a paradox. In his book '*Behave*', he explains that while consciousness allows us to plan for the future and make moral choices, it also brings more anxiety and stress. We can worry about what might happen or feel guilty about our past actions. This stress might seem like a disadvantage for survival. However, Sapolsky suggests that this anxiety is just a side effect of our advanced brain functions, which overall makes us more adaptable and better at surviving despite the added stress.

Richard Dawkins, in *The Selfish Gene* (1976), expressed this challenge, calling consciousness "the most profound mystery facing modern biology." Philosophers like Pascal and Rousseau also remarked on the tension between logic and emotion, reinforcing the mystery of consciousness. Pascal opined, "The heart has its own reasons, which the head can never understand." Rousseau wrote, "Above the logic of the head is the feeling in the heart."[58]

Thomas Nagel's seminal essay, "What Is It Like to Be a Bat?", argues that consciousness is inherently subjective and cannot be fully explained by reductionist or physicalist accounts. He posits that every conscious being possesses a unique, subjective experience that is inaccessible to others. For instance, we can never truly grasp what it's like to be a bat, despite understanding its physiology completely. This essay is pivotal in discussions about qualia, the subjective aspects of experience, and the 'hard problem' of consciousness, which explores the gap between objective physical processes and subjective conscious experience.[180]

This enduring "mind-body" problem has long been a focus of disciplines like philosophy, biology, and psychology, yet no definitive explanation exists. The question of why emotions evolve or how they offer survival value has also perplexed researchers. While physics suggests that physical effects are caused by preceding physical events, human experience often feels as if mental events trigger physical effects. Emotions like fear or pain might have evolutionary advantages, but consciousness remains a deeper mystery.

Neuroscientists consider the mind a by-product of brain activity, a view called 'Epiphenomenalism'. David Chalmers, an Australian philosopher, coined the term ‘The hard problem of consciousness’, referring to the difficulty of explaining how physical processes in the brain produce subjective experiences. He proposed that consciousness maybe an intrinsic property of certain information-processing systems. According to him, if we create an artificial system complex enough to simulate the neural networks of the brain, that system might have conscious experiences. He calls this 'artificial consciousness' or 'machine consciousness.[33, 34]

American philosopher and cognitive scientist Daniel Dennett, on the other hand, takes a more materialistic approach, suggesting that consciousness is not as mysterious as some might think. For him, the brain is essentially a biological machine, and replicating its computational aspects could give rise to consciousness in artificial systems. Similarly, One ancient perspective on consciousness, panpsychism, suggests that all matter, living and non-living, possesses some form of consciousness. This idea, rooted in the philosophies of Thales, Plato, and Spinoza, as well as Eastern traditions like Vedanta and Buddhism, has regained attention in the light of the "hard problem" debate. Vedanta, for instance, explores the connection between the soul (atman), God (brahman), and the world, with schools of thought ranging from complete unity (Advaita) to duality (Dvaita).

Daniel Dennett, believed that once we fully understand consciousness, the "hard problem" will dissolve. Others, like Peter Hacker, dismissed the field of 'consciousness studies' as misguided, arguing that the questions being asked are incoherent and that we must rethink the problem from the ground up. British mathematician-turned philosopher Alfred North Whitehead, known for his process philosophy, viewed reality as a series of interconnected events, with consciousness being an intrinsic property of creation.[7, 53, 80, 167, 197]

In Neuroscience, consciousness is thought to be linked to "neuronal correlates of consciousness" (NCC) a specific set of neuronal activities associated with conscious perception. Technological advances have allowed scientists to manipulate these neuronal correlates in animals and humans, bringing us closer to understanding consciousness. Thomas Huxley defended the view that consciousness is a by-product of neural activity, while William James proposed that consciousness offers survival advantages through natural selection. Sir Karl Popper supported this idea in his book '*The Self and Its Brain.*' However, the controversy about the relationship between brain activity and consciousness continues.[254]

Christof Koch explores the neuronal correlates of consciousness and how technological advances are helping us study conscious perception. Francis Crick connects consciousness with brain activity.[37, 47]

Richard Frackowaik and seven other neuroscientists, in the chapter 'The Neural Correlates of Consciousness' of their book 'Human Brain Function', wrote: "We have no idea how consciousness emerges from the physical activity of the brain, and we do not know whether consciousness can emerge from non-biological systems, such as computers. At this point, the reader will expect to find a careful and precise definition of consciousness. You will be disappointed. Consciousness has not yet become a scientific term that can be defined in this way. Currently, we all use the term consciousness in many different and often ambiguous ways. Precise definitions of different aspects of consciousness may emerge, but to make precise definitions at this stage is premature."[217, 277]

The human mind comprises multiple layers, each with evolutionary significance. The lowest layer encompasses the brainstem and reptilian brain, which regulate fundamental biological functions like heart rate, breathing, eating, and sleeping, alongside primitive instincts. Above this layer lies the limbic system, often referred to as the emotional brain. Between the reptilian brain and the rational brain is the limbic system. It governs emotions and memories. The rational brain resides in the cerebral cortex, the most advanced part of the brain and the last to evolve. It can be said that 'the limbic system proposes and the cortex disposes.' Conscience is not fixed but develops over time, influenced by upbringing, culture, education, and life experiences. When an action contradicts one's moral beliefs, the conscience triggers feelings of guilt, prompting self-reflection and, ideally, behavioural change. Conscience helps individuals navigate complex ethical dilemmas by weighing choices and outcomes based on their values and principles. A well-developed conscience fosters a sense of responsibility towards others, motivating pro-social behaviours like kindness, fairness, and justice.

In summary, while the mind's connection to the brain and the nature of consciousness continue to challenge scientists and philosophers, the debate is far from settled. From Descartes 'dualism' to Chalmers 'hard problem' to modern Neuroscience, the puzzle of how subjective experience arises from physical processes remains one of the great mysteries of human understanding.[48, 60, 84, 141]

Superintelligent AI could have a profound impact on the job market. Many existing jobs may become obsolete, replaced by AI systems capable of automating a wide range of tasks. However, some argue that as technology evolves, new opportunities will arise, and humans will adapt to this transformed reality.

Autonomous systems, ranging from self-driving cars to robotic surgeons, present significant ethical dilemmas. The challenge lies in determining how machines can be programmed to make ethical choices. Should AI be guided by a utilitarian approach, aiming to maximise overall benefits? Or should it prioritise humanistic principles that value individual rights? It is crucial that a machine's moral compass is transparent, impartial, and in line with the evolving ethical standards of society. Determining who defines these standards and how they evolve will continue to be essential questions.

The safe development of AI is important. We need to create "Friendly AI" that aligns with human interests. Public awareness is crucial too. Concentrating AI control in a few hands could worsen inequality and widen the rich-poor gap. To prevent AI from becoming a threat, we should promote responsible and ethical AI development.

As we near the Singularity, we must discuss the ethical and societal impact of Superintelligence.[184, 242]

Mind Uploading and Super intelligent Robotic Mind

"The fabric of the world is mind-stuff."

Arthur Eddington

At the tipping point of technological advancement, we are approaching the possibility of uploading our minds/ brains into robotic bodies of any form-whether human or octopus. Then, the mind can stay as purely digital, having the capacity to have any number of copies and travel at the speed of light. We already know that we can operate machines wirelessly from the other side of the planet. Functional MRI, PET scans, and other technologies can map the brain and its intricate connections, though the challenge is enormous, given the brain's 86 billion neurons and roughly 100 trillion neural connections.[274]

The human brain is like a highly advanced computer, and its intelligence comes from layers of neurons (nerve cells) working together. These layers process information step by step, allowing us to recognise faces, understand language, and decide.

How do we recognise a cat? The bottom layer detects simple features, like edges and shapes. Middle layers combine these features into patterns such as eyes, whiskers, ears, etc. Higher layers recognise the full cat and link it to our memories. Each layer refines the information before passing it to the next, making our recognition smooth and automatic.

Artificial intelligence, especially deep learning, tries to mimic this layered structure. AI models called 'neural networks' have artificial "neurons" arranged in layers. The more layers AI has, the smarter it gets. While AI is inspired by the brain, it's still far from matching human intelligence. However, its ability to learn from data and recognise complex patterns has revolutionised fields like medicine, self-driving cars, and language translation.

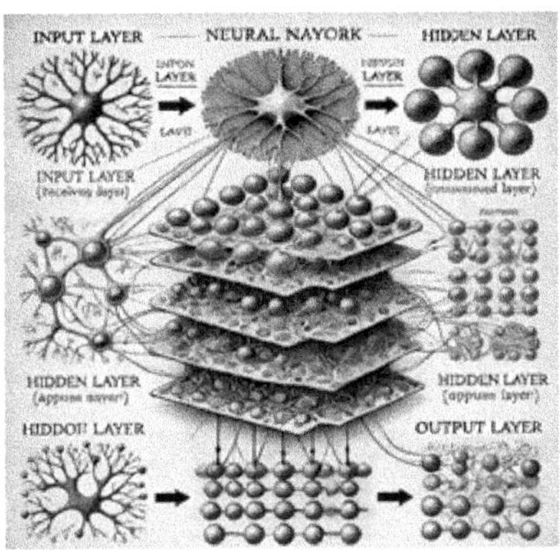

Fig 40: *Neural layers and network. Picture developed by AI*

The expansion of the brain during biological evolution, has been accompanied by a rise in the number of neocortical layers. Mammals have six layers, while reptiles have three and marsupials have four or five. If machines surpass this layers to seven or eight the scenario will be totally different. The development of artificial neural networks (ANNs) or neuromorphic computing systems exceeding the six-layered mammalian cortex, to seven, eight or more layers, may yield transformative effects on artificial intelligence, cognition, and consciousness. This could become different from any biological mind. It may have synthetic consciousness, alien logic, hyper-dimensional awareness and self-generated forms of creativity. It could mean deeper abstraction and meta-thinking, It might represent the next step in cognitive evolution.[215]

Once Peter's mind is uploaded to a robot, the robot would undeniably say, "I am Peter." Can anyone argue this robot is not Peter? If 'Robot Peter' possesses superintelligence, he could assimilate knowledge at a rate exponentially faster than the human Peter ever could. 'Robot Peter' could communicate with other robots and even form communities within the metaverse or real world. They could take over the world in no time, reproducing like copying machines. They could enhance their capabilities by adding more neural layers to their digital systems and travel at the speed of light. In this scenario, robots may dominate the world, and the fate of humans is uncertain. Will we become mere pets or slaves? Or face complete extinction?[248]

The future of ethics and morality will involve grappling with the unprecedented challenges posed by advancing technologies, environmental crises, and deepening global interconnectivity. To navigate these challenges, societies must establish robust ethical frame works that ensure equity, inclusivity, and sustainability.

At this point, we may be reminded of the concept of 'Superman (Übermensch)' as envisioned by Friedrich Nietzsche. Superman represents an idealised future human being who transcends conventional morality and societal norms to create their own values and purpose. Nietzsche introduced this in his work 'Thus Spake Zarathustra' as a symbol of self-overcoming, strength, and the will to power. The 'Übermensch' rejects the limitations of traditional religious and ethical systems, seeking to rise above them through personal transformation and mastery over oneself. Nietzsche's Superman is a metaphor for human potential, embodying creativity, courage, and independence in the face of life's inherent chaos and meaninglessness. The Übermensch is thus an aspirational figure for Nietzsche, representing the pinnacle of human evolution and the ultimate fulfilment of human capability. He questioned, "Man is something that shall be overcome. What have you done to overcome him? All beings so far have created something beyond themselves; and do you want to be the ebb of this great tide and return to the animals rather than overcome man?"[134, 185, 235]

Sri Aurobindo was a revolutionary poet, yogi, and philosopher. His concept of the evolution of consciousness is a transformative vision in which he proposed that life and consciousness evolve continuously, ultimately leading to a higher, divine state. He believed that humanity is not the final step in evolution, and that a new, divine consciousness would eventually emerge, fundamentally transforming human nature.[225, 23]

Immortal Universal species

"The real risk with AI isn't malice but competence."

Nick Bostrom

It is still uncertain whether AI will possess consciousness and conscience, if it reaches or surpasses human intelligence. If AI advances, its priorities may differ from those of humans. In addition, the rise of multiple AIs could lead to competition or collaboration among them, diminishing human dominance.[180]

We still haven't fully deciphered consciousness and conscience. Therefore, we don't know what type of consciousness or conscience 'Robot Peter' would possess. We are aware of the vast disparity of conscience and consciousness between a gorilla and a human. Similarly, there could be a similar gap between humans and super-intelligent robots. Whether robots would have a conscience is debatable. Even if robots have a conscience, it might not connect to sensations like pain, emotions such as fear, love or hate or human values such as empathy, sympathy, and ethics. Perhaps our only solution is to merge with machines, become cyborgs, and gradually shed our biological identities, eventually transforming into fully digital beings capable of transcending the universe at the speed of light. We may have to accept this as the next stage of 'Global Evolution' -becoming an immortal, universal species, a digital intelligent species. Should we call it '*Machina intelligens Meaning, intelligent Machine*' [25,47, 142]

REFERENCES AND BIBLIOGRAPHY

1. Abad M, Mosteiro L, Pantoja C, Cañamero M, Rayon T, Ors I, et al.: Reprogramming in vivo produces teratomas and iPS cells with totipotency features. *Nature*. 2013;502:3405.doi: 10.1038/nature 12586. [PubMed]
2. Adam Bohr and Kaveh Memarzadeh: "Artificial Intelligence in Healthcare"
3. Adobe: "Creative AI: How Adobe Sensei is Changing the Game," Adobe Blog, 2024
4. Adventures of Ideas" (1933)A more accessible exploration of Whitehead's philosophy, connecting his metaphysical ideas with cultural and intellectual history.
5. Alan H. Handyside, Joyce C. Harper, Gedis Grudzinskas "Preimplantation genetic diagnosis and screening: Current status and future challenges" Published in: Human Reproduction Update Volume: 6, Issue 5, Year: 2000 Pages: 395-413
6. Alberts, B., Johnson, A., Lewis, J., Raff, M., Roberts, K., & Walter, P. (2002). Molecular biology of the cell. Garland Science.

7. Alfred North Whitehead:"Process and Reality" (1929)

This foundational text in process philosophy views reality as a dynamic process and consciousness as integral to the unfolding of events in the universe.Read more about it here.

8. American Society of Plastic Surgeons (ASPS). (2022). 2021 Plastic Surgery Statistics Report. https://www.plasticsurgery.org

9. Andersson, M. (1994). Sexual Selection. Princeton University Press. *Research Articles:*

10. Andrew Ng's Machine Learning Course on Coursera

11. Anthony Atala "Stem Cells in Regenerative Medicine: Principles and Prospects" Journal: Science Year: 2011 Volume: 231 Issue: 6020 Pages: 713-720

12. Aurélien Géron.: "Hands-On Machine Learning with Scikit-Learn, Keras, and TensorFlow"

13. Azuma, R. T. (1997). "A Survey of Augmented Reality." *Presence: Teleoperators and Virtual Environments*, 6(4), 355-385.

14. Beevor, A. (2002). *Berlin: The Downfall 1945.* Viking Press.

15. Benefits and Costs of a Big Brain." *Smithsonian Institution's Human Origins Program*. Last updated January 2024. https://humanorigins.si.edu . Human Origins

16. Berger, M. (2016). *Nanotechnology: The Future is Tiny*. Royal Society of Chemistry.
17. Best Practice & Research Clinical Obstetrics & …, 2017
18. Bhushan, B. (2007). *Nanotechnology: Nanomechanics and Nanotribology*. Springer.
19. Big Bang Theory: Evolution of Our Universe.": https://www.universetoday.com
20. Billinghurst, M., Clark, A., & Lee, G. (2015). "A Survey of Augmented Reality." *Foundations and Trends® in Human–Computer Interaction*, 8(2-3), 73-272..
21. Blackburn, E. H., & Epel, E. S. (2012). Telomeres and adversity: Too toxic to ignore. Nature, 490(7419), 169-177.
22. Blum, V. L. (2003). *Flesh Wounds: The Culture of Cosmetic Surgery.* University of California Press.
23. Boisseau, P., & Loubaton, B. (2011). "Nanomedicine, nanotechnology in medicine." *Comptes Rendus Physique*, 12(7), 620–636.
24. Bonetti, F., Warnaby, G., & Quinn, L. (2018). "Augmented reality and virtual reality in physical and online retailing: A review, synthesis and research agenda." *Futures*, 88, 155-169..
25. Bostrom, N. (2014). Superintelligence: Paths, Dangers, Strategies. Oxford University Press. This

book explores the potential risks and benefits of superintelligent AI and discusses strategies for managing these risks.

26. Bowlby, J. (1969). Attachment and Loss: Vol. 1 Attachment. Basic Books.
27. Brian Sykes *Human Evolution: A Visual Guide*
28. Broad Institute CRISPR website: The Broad Institute has been at the forefront of CRISPR research. Their website (www.broadinstitute.org/what-we-do/areas-of-focus/genome-biology/crispr) provides information and resources on CRISPR technology.
29. Buss, D. M. (1994). The Evolution of Desire: Strategies of Human Mating. Basic Books. The book investigates the strategies men and women have developed over time to attract and retain mates, shaped by evolutionary pressures
30. Camus, A. (1942). *Le Mythe de Sisyphe* [*The Myth of Sisyphus*]. Gallimard. Albert Camus's Philosophy of the Absurd:
31. Carroll, R. L. (2001). Patterns and Processes of Vertebrate Evolution. Cambridge University Press.
32. Chaisson, Eric. *Cosmic Evolution: From Big Bang to Humankind.* Center for Astrophysics, Harvard University. Available at:
33. Chalmers, D. J. (1996). The Conscious Mind: In Search of a Fundamental Theory. Oxford University

Press. This book explores the hard problem of consciousness and proposes various theories.

34. Chalmers, D. J. *The Character of Consciousness* (2010): This book compiles Chalmers' essays on consciousness, delving into topics like the nature of experience, the concept of a "zombie," and the foundations of the philosophy of mind.
35. Charles Darwin: The Descent of Man, and Selection in Relation to Sex
36. Christian Boch et al. "Transcription activator-like effectors: a new platform for genome engineering" (2009), *Nature Biotechnology*:
37. Christof Koch – "The Quest for Consciousness: A Neurobiological Approach" (2004)
38. Christopher Bishop: "Pattern Recognition and Machine Learning"
39. Christopher Bishop: : Neural networks for pattern recognition Oxford university press
40. Cohen S.N., Chang A.C., Boyer H.W., Helling R.B. Construction of biologically functional bacterial plasmids in vitro. *Proc Natl Acad Sci,* 70(11):3240-4. 1973. [PubMed]
41. Collins, F. S., & Venter, J. C. (2001). *The Human Genome Project and beyond.* Nature, 409(6822), 860-921.:

42. Collins, F. S., Morgan, M., & Patrinos, A. (2003). *The Human Genome Project: Lessons from large-scale biology.* Science, 300(5617), 286-290.:This article provides an overview of the Human Genome Project, including its goals, challenges, and achievements. It also reflects on the lessons learned from this large-scale biological endeavor.

43. Consciousness and the Brain" by Stanislas Dehaene (2014)

    Dehaene discusses how modern neuroscience has made strides in identifying the neural basis of consciousness and its implications for artificial intelligence.

44. Coursera: "AI-Powered Learning Platforms," Coursera Blog, 2024.

45. Cox TC. (2019). Taking the mouse genome apart: implications for vertebrate biology. Nature Reviews Genetics, 20(11), 645-661.

46. Crespi, B.J. & Semeniuk, C. "Adaptation and Constraint in the Evolution of Reproductive Behaviour in Cetaceans. "*Animal Behaviour*, 67(4), 689–702.

47. Crick, F. (1994). The Astonishing Hypothesis: The Scientific Search for the Soul. Scribner. This book explores the neural basis of consciousness from a scientific perspective.

48. Crick, F. "A Framework for Consciousness"

49. Crick F, Koch C, The claustrum and consciousness: an update
50. D Wang · 2019 ·Gene Delivery to Nonhuman Primate Preimplantation ...
51. Dafni, A. (1984). "Mimicry and Deception in Pollination." *Annual Review of Ecology and Systematics*, 15, 259-278.
52. Daley, J. (2017). At 1.6 Billion Years Old, These Fossils Could Be the Oldest Complex Life. *Smithsonian Magazine*. Retrieved from https://www.smithsonianmag.com
53. Daniel Dennett:"Consciousness Explained" (1991); Dennett's seminal work argues that consciousness is not a mysterious phenomenon but rather can be understood through cognitive science and philosophy.
54. Darwin, C. (1859). On the Origin of Species by Means of Natural Selection, or the Preservation of Favoured Races in the Struggle for [1] Life. This seminal work laid the foundation for the modern theory of evolution.
55. Darwin, C. (1871). The Descent of Man, and Selection in Relation to Sex. John Murray.
56. Darwin, Charles. *Letter to Joseph Dalton Hooker*. 1871.
57. David Cyranoski, Heidi Ledford Published The CRISPR-baby scandal: what's next for human gene-

editing' Authors: Nature Volume: 566 Issue: 7745 Year: 2019 Pages: 440-442

58. Dawkins, R. (1982). The Extended Phenotype: The Gene as the Unit of Selection. Oxford University Press.

59. De Bie FR, Davey MG, Larson AC, et al."Artificial placenta and womb technology: past, current, and future challenges towards clinical translation (2021), *Prenat Diagn*: This review provides a comprehensive overview of the history, current status, and future directions of artificial womb technology, including both artificial placenta and whole womb systems. It's a great starting point for understanding the complexities of this field.

60. Descartes, R. (1641). Meditations on First Philosophy. This philosophical work explores the nature of consciousness and the mind-body problem.

61. Dobell, C. *Antony van Leeuwenhoek and His 'Little Animals': Being Some Account of the Father of Protozoology and Bacteriology*.

62. Drexler, K. E. (1986). *Engines of Creation: The Coming Era of Nanotechnology*. Anchor Books.

63. Dunbar, R. I. M. (1996). Grooming, Gossip, and the Evolution of Language. Harvard University Press.

64. E.J. Kalfoglou, J. Scott Roberts "Preimplantation genetic diagnosis: An overview of socio-ethical and

legal considerations" Published in: Clinical Genetics Volume: 64, Issue 4 Year: 2003 Pages: 315-326

65. Edward J. Larson: The History of Evolutionary Thought" : Provides a comprehensive overview of the development of evolutionary ideas, including de Vries's contribution.
66. Elena A P al: Size of the Human Proteome: The Width and Depth; Int J Anal Chem. 2016; 2016:
67. Encyclopaedia Britannica. *Anaximander*. Available at: Britannica - Anaximander
68. Encyclopaedia Britannica. *Great Chain of Being*. Available at: Britannica - Great Chain of Being
69. Encyclopaedia Britannica. *Spontaneous Generation*. Available at: Britannica - Spontaneous
70. Enquist, M., & Arak, A. (1994). "Symmetry, Beauty, and Evolution". *Nature*, 372(6502), 169–172.
71. Eric Topol: "Deep Medicine: How Artificial Intelligence Can Make Healthcare Human Again"
72. Ernest Cline: Novel "Ready Player One" by (2011) Wired Magazine Articles on the Metaverse
73. Etcoff, N. (1999). *Survival of the Prettiest: The Science of Beauty*. New York: Double day.This source discusses how the golden ratio (1:1.618) is linked to perceptions of beauty in art, nature, and human faces.
74. European Union: "GDPR and AI: What You Need to Know," EU Official, 2023.

75. Evolution: The History of an Idea" by Peter J. Bowler: Another excellent resource tracing the history of evolutionary thought, with a section on de Vries and his mutation theory

76. Faegri, K., & Van der Pijl, L. (1979). *The Principles of Pollination Ecology*. Pergamon Press.

77. Farahany, N. (2004). The ethics of brain enhancement. Nature, 431(7007), 499-503

78. Francis Crick The Astonishing Hypothesis; (1994) Crick, one of the co-discoverers of DNA, connects consciousness with brain activity, emphasizing the importance of the NCC in understanding the mind.

79. Freud, S. (1914). On Narcissism: An Introduction.

80. From Bacteria to Bach and Back: The Evolution of Minds" (2017); This book expands on Dennett's views on the evolution of consciousness and its grounding in natural processes.

81. Futuyma, D. J. (2017). Evolution. Sinauer Associates. This comprehensive textbook provides a modern perspective on evolutionary biology, including the mechanisms of evolution and the evidence supporting it.

82. Galeon, Dom. "Human Evolution and the Cosmos." Futurism. Discusses how cosmic evolution emphasizes increasing complexity over time, highlighting humanity's role as a product of natural processes. Available at: Futurism.

83. Gallup, G. G., & Frederick, D. A. (2010). *The science of sex appeal: An evolutionary perspective.* Review of General Psychology, 14(3), 240-250.

84. Gazzaniga, M. S. (2011). The Cognitive Neurosciences. MIT Press. This comprehensive textbook covers various aspects of cognitive neuroscience, including consciousness, attention, and memory.

85. Gazzaniga, M. S. *Right hemisphere language following brain bisection: a 20-year perspective.* [Review]. Am Psychol. 1983. ;. 38. :.

86. Geertz, C. (1973). The Interpretation of Cultures. Basic Books.

87. Gest, H. "The Remarkable Vision of Antonie van Leeuwenhoek."

    Perspectives in Biology and Medicine, 46(4), 522–527, 2003.

88. Gibson, D.G., Glass, J.I., Lartigue, C., Noskov, V.N., Chuang, R.Y., Algire, M.A., ... & Venter, "Creation of a bacterial cell controlled by a chemically synthesized genome" J.C. Journal: Science Year: 2010 Volume: 329 Issue: 5987 Pages: 52-56

89. Golden Ratio and Beauty. (n.d.). Retrieved from https://www.goldennumber.net: This source discusses how the golden ratio (1:1.618) is linked to perceptions of beauty in art, nature, and human faces.

90. Goodfellow, I., Bengio, Y., & Courville, A. (2016). Deep Learning. MIT Press.

91. Google: "Quantum Supremacy and AI," Google Research, 2023.

92. Gould, S. J. (1977). *Ontogeny and Phylogeny*. Harvard University Press..

93. Gould, S. J. (1989). *Wonderful Life: The Burgess Shale and the Nature of History*. W. W. Norton & Company. Stephen Jay Gould's Concept of "Replaying the Tape of Life

94. Haiken, E. (1997). *Venus Envy: A History of Cosmetic Surgery.* Johns Hopkins University Press.

95. Haldane, J.B.S. "The Origin of Life." *The Rationalist Annual*, 1929.

96. Hall, J. (2007). *A History of Ideas and Images in Italian Art*. Harper & Row. (Discusses the cultural significance of Zeuxis' approach todepicting Helen.)

97. Hannah Holmes- '*The well dressed Ape*':

98. Harley, C. B., Futcher, A. B., & Greider, C.W.(1990).*Telomeres shorten during ageing of human*

99. *Harifield,M.(2003).The evolution of sexual reproduction: a review of theories.Molecular Reproduction and Development,65(4),497-503.*

100. Hennessy,J. L., & Patterson,D.A. (2017).

Computer Architecture: A Quantitative Approach.

101. Henry T Greely, CRISPR'd babies: human germline genome editing in the 'He Jiankui affair'* *Journal of Law and the Biosciences*, Volume 6, Issue 1, October 2019, Pages 111–183

102. Higley, L. G., & Pedigo, L. P. (2012). Economic, ecological, food safety, and social consequences of the deployment of Bt transgenic plants. Annual review of entomology, 57, 153-178.

103. Hooton V, Romanis EC. (2022), *J Law Biosci*: "Artificial womb technology, pregnancy, and EU employment rights

104. https://lweb.cfa.harvard.edu/~ejchaisson/cosmic_evolution/docs/splash.html.

105. https://www.nobelprize.org/prizes/medicine/1959/summary/

106. https://www.universetoday.com; "Big Bang Theory: Evolution of Our Universe." :

107. Human Evolution: Evolution of Brain Size and Adaptations." *Encyclopaedia Britannica*. https://www.britannica.com/science/human-evolution Encyclopedia Britannica

108. Human Rights Watch. (Annual Reports on Global Ethnic Conflicts). www.hrw.org.

109. Hume, D. (1739). A Treatise of Human Nature. Hume's skepticism about the nature of consciousness and the self has influenced subsequent philosophical thought.

110. Huntley, H. E. (1970). The Divine Proportion: A Study in Mathematical Beauty. Dover Publications..

111. Hutchison, C. A., Chuang, R. Y., Noskov, V. N., et al. (2016). Design and synthesis of a minimal bacterial genome. Science, 351(6280), aad6253.

112. Ian Goodfellow, Yoshua Bengio, and Aaron Courville: "Deep Learning"

113. IBM Watson Health

114. Intel. "Excerpts from 'A Conversation with Gordon Moore: Moore's Law."

115. International Human Genome Sequencing Consortium: "Initial sequencing and analysis of the human genome" Journal: Nature Year: 2001 Volume: 409

116. Islam M S: Assessing Nitrogen-Fixing Bacteria and Mineral Nitrogen Fertilization Regimes for Boosting Growth, Photosynthesis, and Essential Oil Production of Clary Sage

117. Islam, M. S., & Shukla, R. (2021). "Green Nanotechnology: Challenges and Opportunities." *Frontiers in Nanotechnology*, 3.

118. J Traeger-Synodinos Pre-implantation genetic diagnosis, Best Practice & Research Clinical Obstetrics & …, 2017

119. Jablonski, N. G. (2012). Skin: A Natural History. Penguin Books..

120. Jablonski, N. G., & Chaplin, G. (2000). The evolution of human skin coloration. Journal of Human Evolution, 39(1), 57-106.

121. Jackson D.A., Symons R.H., Berg P. Biochemical method for inserting new genetic information into DNA of Simian Virus 40:. *Proc Natl Acad Sci,* 69(10): 2904-9 1972. [PubMed]

122. James D. Watson and Francis H. Crick, "A Structure for Deoxyribose Nucleic Acid," Nature, Vol. 171, No. 4356, pp. 737-738, April 1953.

123. Jens Boch and Ulla Bonas :"TAL effectors: customizable proteins for DNA targeting" (2010), *Nature Reviews Genetics*: This review article

124. Jerald, J. (2015). *The VR Book: Human-Centered Design for Virtual Reality*. ACM Books

125. Jeremy D. Kark, Abraham Aviv; "Telomere biology and human : A causal link with cardiovascular ageing?" American Journal of Human Biology Volume: 16, Issue 5 Year: 2004 Pages: 641-650

126. Johanson, D. C., & Shreeve, J. (1989). Lucy: The Beginnings of Humankind.

127. Johanson, D. C., & White, T. D. (1979). A systematic and evolutionary perspective on the African early hominid material. Science, 203(4379), 321-330.

128. Johns Hopkins University Press published a translated edition of *De revolutionibus* (E. Rosen, 1992).

129. Jones, D., & Hill, K. (1993). "Criteria of Facial Attractiveness in Five Populations". *Human Nature*, 4(3), 271-296.:

130. JRV e r meesch,T Voet,K Devriend t Prenatal and pre-implantation genetic diagnosis- Nature Reviews Genetics, 2016

131. kanazawa,S (2011). Intelligence aand physical attractiveness: Indicators of genetic fitness.*Personality and Individual Differences*, 50 (5),733,737

132. Karl Popper and John Eccles "The Self and its Barin"(1977)

133. Kathleen Mc Auliffe (https: www. discover magazine.com/mind/if-modern-humans-are-so-smart-why-are-our-brains-shrinking)

134. Kaufmann, W. A. (1968). Nietzsche: Philosopher, Psychologist, Antichrist. Princeton University PressThis classic work offers a comprehensive

135. Keane TM,Goodstadt L.Danecek P,White MA,wong K,Yalcin B,etal.(2011) Mouse genomic variation and its effect on phenotypes and gene regulation Nature,477(7364),289-294.

136. Keats J. (1818)

Endymion:A poetic Romance Book I

137. Kimbel, W. H., Johanson, D. C., & Rak, Y. (1994). Systematic assessment of early hominid remains. Science, 264(5163), 1335-1345.

138. Kirkpatrick, M. (1982). Sexual selection and the evolution of female choice. Evolution, 36(1),1-12.

139. Kissinger, H. (1972). As quoted in *The New York Times*, January 19, 1972, by journalist Tom Braden

140. Klug, W. S ,Cummings,M .R. ,Spencer, C .A , & Pallagina,M.A(2012).

141. Koch, C. (2004). The Quest for Consciousness:

A Neurobiological Approach. Roberts & CompanyPublishers. This book delves into the neuroscience of consciousness, discussing the networks.

142. Kurzweil, R. (2005). The Singularity Is Near: When Humans Transcend Biology. Viking. This book presents a future in which technological

technological singularity.

143. Lander ES, Linton LM, Birren B, Nusbaum C, Zody MC, Baldwin J, et al. (2001). Initial sequencing and analysis of the human genome. Nature, 409(6822), 860-921

144. Lander, E. S. (2015). "The Heroes of CRISPR." *Cell*, 164(1-2), 18-28. DOI:10.1016/ j.cell.2015.12.041.

145. Langlois, J. H., Kalakanis, L., Rubenstein, A. J., & Langlois, A. C. (2000). Attractive faces are prototypical: Infants prefer average faces. Developmental Psychology, 36(2), 223-230.

146. Lartigue, C., Vashee, S., Algire, M. A., et al. (2009). Creating bacterial strains from genomes that have been cloned and engineered in yeast. Science, 325(5948), 1693-1696.

147. Laura Mulvey: The Male Gaze by: This essay explores the role of male gaze in shaping female representation in media and culture.

148. Lazzaro Spallanzani's Contributions to Reproduction Biology *"The Emergence of Embryology in the Early Modern Period"*, Springer, 2014..

149. Le Cong et al."RNA-programmed genome editing in human cells" (2013), *Science*: This paper showed the application of CRISPR-Cas9 for genome editing in human cells.

150. LeCun, Y., Bengio, Y., & Hinton, G. (2015). Deep learning. Nature, 521(7553), 436-444.

151. Lee, K. (2012). "Augmented reality in education and training." *TechTrends*, 56(2), 13-21.

152. Lee, L.K., Cheung, S.K.S., & Kwok, L.F. Learning Analytics: Current Trends and Innovative Practices

153. Lewin, R. (2006). Human Evolution: An Illustrated Introduction. Blackwell Publishing

154. Livio, M. (2002). The Golden Ratio: The Story of Phi, the World's Most Astonishing Number. Broadway Books.

155. Lovejoy, C. O. (1981). *The Origin of Man.* Science, 211(4480), 341-350.

156. Lovejoy, C. O., Latimer, B., Suwa, G., White, T. D., & Asfaw, B. (1999). The pelvis and femur of Australopithecus afarensis. Science, 286(5448), 1335-1339.

157. Luca Pacioli's *De Divina Proportione* (1509):

158. Lyons, T. W., Reinhard, C. T., & Planavsky, N. J. (2014). The rise of oxygen in Earth's early ocean and atmosphere. *Nature*, 506 (7488), 307–315. https://doi.org/10.1038/nature13068

159. Maienschein, J. *"Preformation or Epigenesis? Historical Discussions of Reproductive Biology." Developmental Biology*, 112(1), 1–11, 1985.

160. Maienschein, J. *"Spallanzani and the Question of Spontaneous Generation."* In *Biology and the Foundations of Ethics* (eds. J. Maienschein & M. Ruse), Cambridge University Press, 1999..

161. Maiorana, V. C. *"Spallanzani's Contributions to Reproductive Biology." American Zoologist*, 30(4), 837–844, 1990.

162. Mallucci, P., & Branford, O. A. (2012). Concepts in aesthetic breast dimensions: Analysis of the ideal breast. *Journal of Plastic, Reconstructive & Aesthetic Surgery*, *65*(1), 8–16. https://doi.org/10.1016/j.bjps.2011.08.006

163. Mario Livio's *The Golden Ratio: The Story of Phi, the World's Most Astonishing Number* (2002):

164. Martin Jinek et al "A programmable dual-RNA-guided DNA endonuclease in adaptive bacterial immunity". (2012), *Science*:

165. Matthew Ball's "The Metaverse: And How it Will Revolutionize Everything" (2022)

166. Maurice Wilkins, A. R. Stokes, and H. R. Wilson, "Molecular Structure of Deoxypentose Nucleic Acids," Nature, Vol. 171, No. 4356, pp. 738-740, April 1953.

167. Max Velmans and Susan Schneider (Edited by); "The Blackwell Companion to Consciousness"

168. Maxwell Bennett, Daniel Dennett, Peter Hacker, and John SearleNeuroscience and Philosophy: Brain, Mind, and Language" (2007) by

169. Maxwell Bennett, Daniel Dennett, Peter Hacker, and John Searle; Neuroscience and Philosophy: Brain, Mind, and Language" (2007)

170. Mayr, E. (2001). What Evolution Is. Basic Books.

171. McKinsey & Company: "The Future of Autonomous Vehicles," McKinsey, 2023

172. Meissle, M., Romeis, J., & Bigler, F. (2011). Bt maize and integrated pest management—A European perspective. Pest management science, 67(9), 1049-1058.)

173. Meta (formerly Facebook) on the Metaverse

174. Michod, R. E. (1997). Evolution of the sexes: An evolutionary perspective on sexual reproduction. American Naturalist, 150(4), S50-S60

175. Milgram, P., Takemura, H., Utsumi, A., & Kishino, F. (1994). "Augmented Reality: A class of displays on the reality-virtuality continuum." *Proceedings*

of Telemanipulator and Telepresence Technologies.

176. Miller, S. L. (1953). A Production of Amino Acids Under Possible Primitive Earth Conditions. *Science*, 117(3046), 528–529. doi:10.1126/science.117.3046.528.

177. Moore, G. E. (1965). Cramming more components onto integrated circuits. Electronics Magazine, 38(8), 114-117.

178. Morris Goodman *Chimpanzee and Human Evolution*.

179. Müller, S., & Stewart, J. (2021). "CRISPR-Cas Systems: A Journey from Basic Science to Transformative Technology." *Nature Reviews Genetics*, 22(2), 75-91. Available online at: Nature Reviews Genetics.

180. Nagel, T. (1974). What is it like to be a bat? Philosophical Review, 83(4), 435-450..

181. Naomi Wolf: The Beauty Myth: This book critiques societal beauty standards and their impact on women's lives.

182. Neal Stephenson: Novel "Snow Crash" (1992)

183. Neal Stephenson's "Lamina1" Project

184. Nick Bostrom's "Ethical Issues in Advanced Artificial Intelligence" (2003)

185. Nietzsche, F. (1883-1885). Thus Spake Zarathustra. This is the primary source for Nietzsche's concept of the Übermensch.

186. Nietzsche, F. (1887). *Zur Genealogie der Moral* [*On the Genealogy of Morality*]. Friedrich Nietzsche's Exploration of Existential Themes

187. NIH (National Institutes of Health) on AI in Healthcare

188. Nilsson, N. J. (2010). Artificial Intelligence: A New Synthesis. Morgan Kaufmann..

189. NVIDIA: "AI Chips and Autonomous Systems," NVIDIA Press, 2024.

190. Ochman, H., Lawrence, J.G., & Groisman, E.A. "Lateral gene transfer and the nature of bacterial innovation."*Nature*, 405(6784), 299–304.

191. OECD: "OECD AI Principles," OECD, 2023.

192. Oparin, Alexander. *The Origin of Life*. 1936.

193. OpenAI: "GPT-4: Applications in Creative Writing," OpenAI, 2023.

194. Partridge EA, Davey MG, Hornick MA, et al. "An extra-uterine system to physiologically support the extreme premature lamb (2017), *Nat Commun*:

195. Paul Doty, Marshall Nirenberg, Gobind Khorana, Severo Ochoa Journal "Codon Recognition and Anticodon Structure in tRNA: tRNA's Essential

Role in the Expression of Genetic Information": Science Year: 1968 Volume: 160 Issue: 3829

196. Paul, D. B. (1995). *Controlling Human Heredity: 1865 to the Present*. Humanity Books.:

197. Peter Hacker:"The Philosophical Foundations of Neuroscience" (with M.R. Bennett, 2003). Hacker critiques much of contemporary neuroscience and its assumptions about consciousness, arguing that some questions in consciousness studies are incoherent. Interviews and Lectures. Hacker often discusses his views on consciousness and philosophy in public forums. [Search for his interviews on platforms like YouTube or academic journals.]

198. Peter J. Bowler: "Evolution: The History of an Idea" by: Another excellent resource tracing Peter J. Bowler the history of evolutionary thought, with a section on de Vries and his mutation theory.

199. Philadelphia, The Children's Hospital of (28 February 2017). "A Unique Womb-Like Device Could Reduce Mortality and Disability for Extremely Premature Babies")

200. Pinker, S. (1997). How the Mind Works. W. W. Norton & Company..

201. Ponganis, P.J. *Marine Mammal Physiology: Requisites for Ocean Living*.

202. Potts, R. (1998). Variability selection in hominid evolution. *Evolutionary Anthropology*, 7(3), 81-96.

203. Potts, R., & Faith, J. T. (2015). Alternating high and low climate variability: The context of natural selection and speciation in Plio-Pleistocene hominin evolution. *Journal of Human Evolution*, 87, 5-20.

204. Prashant Mali et al "Multiplex Genome Editing Using CRISPR/Cas Systems". (2013), *Science*: This study demonstrated the ability of CRISPR-Cas9 to edit multiple genes simultaneously.

205. Preimplantation genetic diagnosis: an overview of socio-ethical and legal considerations" Authors: E.J. Kalfoglou, J. Scott Roberts Published in: Clinical Genetics Volume: 64, Issue 4 Year: 2003

206. Primordial Soup Theory. BioScience and Wikipedia

207. Proctor, M., Yeo, P., & Lack, A. (1996). The Natural History of Pollination. Timber Press.

208. Proctor, R.N.(1988). Racial Hygiene Medicineunder the Nazils. Harvard University Press.

209. Prokop P.,& frdor P.(2013). Physical attractiveness

210. Puts, D. A., Jones, B. C., & De Bruine, L.M.(2012).

211. Rat Genome Sequencing Project Consortium. Genome sequence of the Brown Norway rat yields insights into mammalian evolution. *Nature* 428, 493–521 (2004).

212. Ratner, M., & Ratner, D. (2003). *Nanotechnology: A Gentle Introduction to the Next Big Idea*. Prentice Hall.

213. Raup, D. M. (1991). Extinction: Bad Genes or Bad Luck? W. W. Norton & Company.

214. Reidenberg, J.S., & Laitman, J.T. "Cetacean Evolution: Adaptations for Aquatic Life."*The Anatomical Record*, 290(6), 507–513. Reproductive success. *Anthropological Review, 76*(2), 203–216.:

215. Research Pod "Are we on the verge of uploading human consciousness?"

    This article explores the potential for brain-machine interfaces to enable consciousness transfer into artificial systems.

216. Rhodes, G. (2006). "The Evolutionary Psychology of Facial Beauty". *Annual Review of Psychology*, 57, 199–226.

217. Richard Frackowiak et al: "The Neural Correlates of Consciousness" (2003), Chapter in *Human Brain Function* by.

218. RJ Jones, MR DeBaun - Leukemia after gene therapy for sickle cell disease: insertional mutagenesis, busulfan, both, or neither. Blood, 2021

219. Robin Marantz Henig: "The Monk in the Garden: The Lost and Found Genius of Gregor Mendel, the Father of Genetics" A well-regarded biography that provides context and explains Mendel's work in an accessible way.

220. Roger Scruton The Measure of Beauty

221. Rosalind Franklin, R. G. Gosling, "Molecular Configuration in Sodium Thymonucleate," Nature, Vol. 171, No. 4356, pp. 740-741, April 1953.

222. Russell, S. J., & Norvig, P. (2016). Artificial Intelligence: A Modern Approach. Pearson..

223. Ryan, M. J., & Keddy-Hector, A. (1992). Sexual selection and sensory systems in the evolution of animal communication. American Naturalist, 139(3), S23-S38.

224. satprem:sri. Aurobindo : or the Adventure of Consciousness

225. Sagan Carl. 'Dragans of Eden' consciousness

226. Satprem:. Life Without Death. A follow up to Mind of the Cells, by Satprem. ...

227. Schiestl, F. P., & Cozzolino, S. (2008). "Evolution of Sexual Mimicry in Orchids." *Biological Reviews*, 83(4), 515-532.

228. Searle, J. R. (1980). Minds, brains, and programs. Behavioral and Brain Sciences, 3(3), 417-457.

229. Semir Zeki: Neuroaesthetics: At the Intersection of Art, Science, and Philosophy

230. Shakespeare, W. (1600). *The Merchant of Venice.* While the famous line "Beauty lies in the eyes of the beholder" is widely attributed to Shakespeare, it does not appear verbatim in *The Merchant of Venice* or any of his works.

231. Shin JS, Hong A, Solomon MJ, Lee CS.Pathology. 2006 Apr;38(2):103-13.

232. Shreeve, J. (2004). *The Genome War: How Craig Venter Tried to Capture the Code of Life and Save the World.* Knopf.:

233. Singh, D. (1993). Adaptive significance of female physical attractiveness: Role of waist-to-hip ratio. *Journal of Personality and Social Psychology, 65*(2), 293–307.

234. Smith, E., & Szathmáry, E. (1995). The major evolutionary transitions. *Nature*, 374(6519), 227–232. https://doi.org/10.1038/374227.

235. Solomon, R. C. (1976). Nietzsche: A Very Short Introduction. Oxford University Press. This book provides a concise introduction to Nietzsche's philosophy, including the concept of the Übermensch.

236. Sri Aurobindo: The Life Divine; The Synthesis of Yoga; Savitri: A Legend and a Symbol.

237. Stanford University: "AI in Precision Medicine," Stanford Medicine, 2023.

238. Stanislas Dehaene Consciousness and the Brain" (2014);

239. Stanley N. Cohen, Annie C. Y. Chang, Herbert W. Boyer, Robert B: Construction of Biologically Functional Bacterial Plasmids In Vitro:. Helling Journal: Proceedings of the National Academy of Sciences of the United States of America (PNAS) Year: 1973 Volume: 70 Issue: 3240-3244.

240. Stecket,R.H., & Prince, J.M.(2001)."Tallest in the World: Native Americans of the Great Plains in the Nineteenth Century." American Economic Review,91(1),287 - 294.

241. Stringer,C.(2011).The Neanderthals.Thames & Hudson.

242. Stuart Russell's "Human Compatible: Artificial Intelligence and the Problem of Control" discusses the necessity for AI systems to be aligned with human values.

243. T Shin JS, Hong A, Solomon MJ, Lee CS he role of telomeres and telomerase in the pathology of human cancer and ageing.

244. Takahashi, K., & Yamanaka: "Induction of Pluripotent Stem Cells from Mouse Embryonic and Adult Fibroblast Cultures by Defined Factors", S. Journal: Cell Year: 2006 Volume: 126 Issue: 4 Pages: 663-676

245. The Astonishing Hypothesis" by Francis Crick (1994). Crick, one of the co-discoverers of DNA, connects consciousness with brain activity, emphasizing the importance of the NCC in understanding the mind.

246. The Blackwell Companion to Consciousness' (Edited by Max Velmans and Susan Schneider). This comprehensive resource includes essays on Dennett, Hacker, and other thinkers, offering various philosophical perspectives on consciousness.

247. The Cambridge Handbook of Consciousness (2007)

This edited volume provides a comprehensive overview of scientific, philosophical, and psychological perspectives on consciousness, including a discussion of NCC and alternative theories.

248. The Future of Humanity Institute; 'Whole Brain Emulation: A Roadmap' This academic report from Oxford University outlines the technical requirements and milestones necessary for achieving whole-brain emulation.

249. The History of Evolutionary Thought' by Edward J. Larson:

250. The National Center for Science Education: This organization promotes science education and defends the teaching of evolution in schools.

251. The Royal Society - Evolutionary Biology Insights: https://royalsociety.org.

252. The Singularity Is Near: When Humans Transcend Biology by Ray Kurzweil:

253. The University of California Museum of Paleontology:.

254. Thomas Huxley's Essays on Consciousness; consciousness as an "epiphenomenon,

255. Thomas, C.M., & Nielsen, K.M. "Mechanisms of, and barriers to, horizontal gene transfer between bacteria."*Nature Reviews Microbiology*, 3(9), 721.

256. Thornhill, R., & Gangestad, S. W. (1999). "Facial Attractiveness". *Trends in Cognitive Sciences*, 3(12), 452-460.

257. Threadgill DW, Yee D, Matin A, Nadeau JH, Magnuson T. (1997). Genealogy of the 129 inbred strains: 129/SvJ is a contaminated inbred strain. Mammalian Genome, 8(6), 390-393.)

258. Tony Parisi : "Learning Virtual Reality" by Google Earth VR Museums Offering Virtual Tours, VR in Training Applications

259. Tony Parisi : "Ready Layer One" (Approximately 2.7 years ago): In this Medium article, Parisi reflects on his journey and the future of building an open metaverse.

260. Turing, A. M. (1950). Computing Machinery and Intelligence. Mind, 49(236), 433-460.

261. University of Oklahoma's resources dedicated to Galileo and the Church.

262. Urey, H. C., & Miller, S. L. (1959). Organic Compound Synthesis on the Primitive Earth. *Science*, 130(3370), 245–251. doi:10.1126/science.130.3370.245.

263. Venter, J. C., Adams, M. D., Myers, E. W., et al. (2001). *The sequence of the human genome.* Science, 291(5507), 1304-1351.:

264. VentureBeat: "The Metaverse and the Vision of Neal Stephenson".

265. Victoria P. Beltrami, Giuseppe Mandraffino, Antonino Tuttolomondo "Telomeres and Telomerase in Cardiovascular Diseases": High Blood Pressure & Cardiovascular Prevention Volume: 23, Issue 1, Year: 2016 Pages: 1-8

266. Vince, J. (2004). *Introduction to Virtual Reality*. Springer.

267. Virtual Reality Society (VRS) Scientific American Articles on VR Oculus (Meta Quest)

268. W Nawaz, B Huang, S Xu, Y Li, L Zhu, H Yiqiao: AAV-mediated in vivo CAR gene therapy for targeting human T-cell leukemia. Blood Cancer Journal volume 11, Article number: 119 (2021)

269. Warren Booth:"Parthenogenesis in Reptiles: An Evolutionary Perspective" Journal: *Reproductive Biology:* 2001

270. Warren Booth:"Parthenogenesis in Vertebrates: Myths and Mechanisms", journal of Integrative Zoology: 2007

271. Weindling, P. (1989). *Health, Race and German Politics between National Unification and Nazism, 1870–1945*. Cambridge University Press.

272. Weintraub, Karen (October 19, 2021). "Pig Kidneys to the Rescue? Groundbreaking Transplant a Step Toward Solving our Organ Shortage". *USA Today*. Retrieved September 11, 2022.

273. Wendorf M. (2022), *Science And Stuff* "Exclusive: Concept Unveiled for the World's First Artificial Womb Facility":

274. Wikipedia; "Mind Uploading"

275. Wikipedia's page on *Dialogue Concerning the Two Chief World Systems*

276. Willett, E. (2008, October 29). *Helen of Troy and Medusa: The Power of Faces in History and Mythology*. https://edwardwillett.com

277. William James – "The Principles of Psychology" (1890).

278. Willmer,P.(2011). Pollination and Floral Ecology. Princet on University Press.

279. Wilson,E.O.(1992).The diversity of life.Harvard University Press.

280. World Economic Forum on AI in Healthcare

281. World Economic Forum: "Future of Jobs Report," WEF, 2025.

282. Wrangham, R., & Peterson, D. (1996). *Demonic Males: Apes and the Origins of Human Violence.* Houghton Mifflin Harcourt.

283. Wu, F., & Butzler, T. M. (2012). Bt corn and European corn borer: long-term success through resistance management. Journal of economic entomology, 105(3), 781-792.

284. Zebrowitz, L. A., & Montepare, J. M. (2008). "Social Psychological Face Perception: Why Appearance Matters". *Social and Personality Psychology Compass*, 2(3), 1497-1517.

❍❍❍

www.ingramcontent.com/pod-product-compliance
Ingram Content Group UK Ltd.
Pitfield, Milton Keynes, MK11 3LW, UK
UKHW021406190726
13845UKWH00028B/912

9 788198 576453